Environment, Trade and Society in Southeast Asia

Verhandelingen van het Koninklijk Instituut voor Taal-, Land- en Volkenkunde

Edited by

VOLUME 300

The titles published in this series are listed at *brill.com/vki*

Environment, Trade and Society in Southeast Asia

A Longue Durée *Perspective*

Edited by

David Henley
Henk Schulte Nordholt

BRILL

LEIDEN | BOSTON

KITLV

The realization of this publication was made possible by the support of KITLV (Royal Netherlands Institute of Southeast Asian and Caribbean Studies).

Cover illustration: Kampong Magetan by J.D. van Herwerden, 1868 (detail, property of KITLV).

Library of Congress Cataloging-in-Publication Data

Environment, trade and society in Southeast Asia : a longue durée perspective / edited by David Henley, Henk Schulte Nordholt.

pages cm. -- (Verhandelingen van het Koninklijk Instituut voor Taal-, Land- en Volkenkunde ; volume 300)

Papers originally presented at a conference in honor of Peter Boomgaard held August 2011 and organized by Koninklijk Instituut voor Taal-, Land- en Volkenkunde.

Includes bibliographical references and index.

ISBN 978-90-04-28804-1 (hardback : alk. paper) -- ISBN 978-90-04-28805-8 (e-book)

1. Southeast Asia--History--Congresses. 2. Southeast Asia--Civilization--Congresses. 3. Southeast Asia--Environmental conditions--History--Congresses. 4. Southeast Asia--Commerce--History--Congresses. I. Henley, David, 1963- editor. II. Schulte Nordholt, Henk, 1953- editor. III. Boomgaard, P., 1946- honouree. IV. Bankoff, Greg. Deep forestry. Container of (work):

DS525.32.E58 2015

959--dc23

2014048006

This publication has been typeset in the multilingual "Brill" typeface. With over 5,100 characters covering Latin, IPA, Greek, and Cyrillic, this typeface is especially suitable for use in the humanities. For more information, please see www.brill.com/brill-typeface.

ISSN 1572-1892

ISBN 978-90-04-28804-1 (hardback)

ISBN 978-90-04-28805-8 (e-book)

This book is printed on acid-free paper.

Contents

Acknowledgements

In August 2011, KITLV organized a workshop to honour professor Peter Boomgaard on the occasion of his retirement. We thank KITLV for financing this workshop on the *longue durée* in Southeast Asian history, which formed the basis of the present book. We thank Ellen Sitinjak and Yayah Siegers for their indispensable assistance in organizing this meeting.

We also express our gratitude to the Royal Netherlands Academy of Arts and Sciences (KNAW) for subsidizing the Open Access publication of this book.

Finally we are grateful to Ireen Hoogenboom for the careful and efficient way in which she edited the present volume.

David Henley and Henk Schulte Nordholt

CHAPTER 1

Introduction

Structures, Cycles, Scratches on Rocks

David Henley and Henk Schulte Nordholt

The purpose of this book is to celebrate the work of Peter Boomgaard, recently retired from the University of Amsterdam and the Royal Institute of Southeast Asian and Caribbean Studies (KITLV) in Leiden. Boomgaard is among the leading historians of Southeast Asia in the *longue durée* – that is, over very long periods of time. He single-handedly pioneered the field of Indonesian environmental history, and is the author of the only comprehensive book to date on the environmental history of Southeast Asia as a whole (Boomgaard 2007b). He has written the standard work on the population history of Java (Boomgaard 1989a), and the only book to date on the history of the relationship between tigers and people anywhere in the world (Boomgaard 2001a). Besides his own books he has edited or co-edited 15 collective volumes, written more than 70 chapters in edited volumes, and published more than 40 journal articles, most of them dealing with environmental history or with other aspects of the *longue durée* in Southeast Asia. His fascination with the long term is illustrated by such characteristic Boomgaard titles as 'Land Rights and the Environment in the Indonesian Archipelago, 800–1950' (Boomgaard 2011), 'Economic Growth in Indonesia, 500–1900' (Boomgaard 1993), and even 'Early Globalization: Cowries as Currency, 1600 BCE–1900' (Boomgaard 2008c).

Boomgaard's eye for the broad scheme of things has led him to some unexpected and trail-blazing conclusions. He showed, for instance, that Java deurbanized under colonial rule, the proportion of its population living in towns falling in the course of the nineteenth century, whereas the opposite had usually been assumed (Boomgaard 1989a:110–116). He definitively disproved the once popular theory that Javanese villages were colonial inventions or creations (Breman 1980; Boomgaard 1991). He pointed out that tigers, in Southeast Asia, were not creatures of the natural forest, but thrived where human action created open landscapes rich in large prey animals like deer and pigs, so that tigers were effectively symbiotic with people and for a long time tiger numbers rose rather than declined as the human population of the region increased (Boomgaard 2001a:22–24). He was among the first to understand the importance of traditional fertility control methods to the demographic history of

 DOI 10.1163/9789004288058_002

Indonesia (Boomgaard 1989a:192–196), and later showed how traditional social institutions such as bridewealth and slavery created incentives for women to practice birth control (Boomgaard 2003a, 2003b). His ongoing work on the histories of medicine and forestry in Southeast Asia may be expected to yield comparably radical and productive insights.

For all his fascination with the 'big picture', Boomgaard has seldom been tempted into abstract or theoretical speculation. On the contrary, a look at his publication list (reproduced at the end of this volume) reveals a trademark interest in the worldly, the earthy, even the seamy and sinister side of Southeast Asian history: roots and tubers, leprosy and syphilis, cockfighting and tiger-baiting, bestiality and incest. Above all, Boomgaard is a richly empirical scholar whose publications are mines of information that others, if they wish, can use to generate and test hypotheses quite different from those which he himself advances in relation to them. We believe that the contributions included in the present volume are written in this same empirical spirit, with attention to detail as well as to the broad sweep of history.

In order to honour Peter's contribution to Southeast Asian history, in 2011 we invited some of his closest colleagues to address the question of what new insights a long-term historical perspective adds to our understanding of Southeast Asia. This book is the end result. The time span covered is from about 800 to 2000 CE: that is, from the time of Borobodur to the present. While few chapters cover the whole of this period, all deal with a time span of at least two centuries, and all are concerned with the historiographic big picture: the identification of processes and events that have shaped and changed the region in lasting ways. The immediate methodological advantage of such a long-term perspective is that it forces us to overcome the compartmentalization imposed by conventional historical periodization, by which each era tends to be portrayed in terms of a different set of analytical concepts coined by a different group of historians claiming expertise in that particular era.

Among Boomgaard's most important sources of intellectual inspiration over the years has been the great French historian Fernand Braudel, by whose work he has described himself as 'deeply influenced' (Noordegraaf 2006:50). It was Braudel and his 'Annales School' who made the term *longue durée* common currency among historians everywhere, and who inspired a grand debate about the causes of historical change which still continues today. Before going on to discuss the historiography of Southeast Asia in the *longue durée* and to preview the contents of this volume, some words are in order about Braudel and the origins of the *longue durée* as a historiographic concept.

Fernand Braudel (1902–1985) is perhaps most famous for his three-volume history and historical geography of *The Mediterranean and the Mediterranean world under Philip II* (1949, 1972–73). In 1958, however, he also distilled his practical experience of writing 'broad sweep' history into a theoretical article on the relationship between structure and time in historical research that was to prove agenda-setting. Entitled 'History and the Social Sciences: The *Longue Durée*' (Braudel 1958, 1960, 1980), this was partly a reaction against the views of the anthropologist Claude Levi-Strauss, who argued in *L'Anthropologie Structurale* (1958) that in order to isolate the fundamental structures that order human life, it is first necessary to eliminate the interference of time and history. Braudel admired Levi-Strauss's structuralist quest for a 'social mathematics', but insisted that no human phenomenon is ultimately independent of historical change.

In elaboration, Braudel proposed a hierarchy of three interlocking historiographic time frames: the 'history of events', the history of cycles, and the *longue durée* or long duration. The history of events corresponded to traditional historiography, describing events from day to day, month to month, and year to year. As an example of cyclic history, Braudel cited the 50-year cycle of rising and falling prices identified in European history by the economist Nikolai Kondratiev. The third time frame, the *longue durée*, Braudel saw mainly in terms of persistent structures embedded in environmental constraints.

> Certain long-lived structures are stable elements for many generations; they encumber history, and by disturbing it they determine its course. [...] As constraints, they are the limitations from which man and his experiences can liberate themselves only with difficulty. [...] Geographical constraint is a ready example. Through the centuries, man has been the prisoner of climatic and other geographical limitations, and of a slowly-achieved equilibrium, from which he can depart only at great risk.
>
> BRAUDEL 1960:6

It was the influence of geographical and environmental factors on the history of the Mediterranean which had been the topic of the first, and in many ways most innovative, volume of Braudel's own magnum opus (1949). Another important type of structural constraint in the *longue durée*, for Braudel, was the momentum of established cultural forms, such as those of the Latin civilization which led the elites of medieval Europe to live for centuries 'by the same themes, comparisons, commonplaces, and slogans' (Braudel 1960:6).

Braudel did not specify to what extent his three time frames were related to each other, or by what kinds of causal connections he believed they were linked. But he clearly believed that the longer the time perspective employed, the greater the chance of interpreting a given event or development correctly. For Braudel, 'short-run time', measured 'by our illusions and fleeting moments of consciousness', was 'the most capricious and deceptive of durations'.

> The surpassing of short-run time seems the outstanding accomplishment of the historiography of the last one hundred years. Thus one can understand the pre-eminent role of the history of institutions, religions, and civilizations; and, thanks to archeology, which covers vast chronological spaces, the importance of studies of classical antiquity.
>
> BRAUDEL 1960:4–5

His main intention, in fact, was to redirect the attention of historians away from the study of political events, which can be as evanescent as newspaper headlines, and toward persistent structures, such as those of culture and geography.

Among historians of Southeast Asia, attention to persistent structures is in itself nothing new. Indeed, historical continuity was something of an obsession among the most influential pioneers of the field in the mid-twentieth century. Still a nice quotation to begin a lecture with is B.J.O. Schrieke's bold declaration (1955–57, II:100) in *Indonesian Sociological Studies* that 'the Java of around 1700 AD was in reality the same as the Java of around 700 AD'. In the colonial period (and to some extent up to now) it was conventional to divide the Indonesian past into an evolutionary sequence of periods according to the presence of external civilizing influences – prehistoric, Indian, Islamic, European. But this, Schrieke argued, was misleading.

> In such studies one is too much under the influence of the dynamism of the familiar modern Western world, whereas the static society [...] cherished no other ideal than to remain as it was, shunning all change. [...] [T]he primary factors which determined the structure of Javanese society remained unchanged. The first of those primary factors was the Javanese people, vis-à-vis of whom the foreign, immigrating elements always formed an insignificant minority [...]. The second factor was the Javanese landscape, which was able to retain its dominating influence through the ages because of the fact that in all those ages [...] there was no technological development making it possible to overcome the obstacles of the geographical situation.
>
> SCHRIEKE 1955–57, II:99–100

In this view, which Schrieke believed was also the view of the Javanese themselves, historical change in the precolonial period had largely been limited to the cyclic rise and fall of transient kingdoms and dynasties.

> The static character of the primary factors has had the result that if history has repeated itself anywhere it has certainly been on Java [...]. [R]epetitions recurred time and again. The Javanese themselves appear to have observed this: in any event (perhaps in connection with Indian *yuga* concepts) the belief arose among them that in each dynasty a number of similar events must needs occur [...].
>
> SCHRIEKE 1955–57, II: 100–101

Like Schrieke (1890–1945), other scholars of his time also sought to relativize the impact of foreign civilizations on Southeast Asian cultures and ways of life. In French Indochina, Paul Mus (1902–1969) argued that the Cham and Vietnamese peoples had assimilated Indian and Chinese religious concepts and practices to their own autochthonous cults of agricultural fertility and ancestor-worship, in which chiefs and their ancestors mediated between the community and its earth-gods (Mus 1933, 2011). Schrieke's colleague Jacob van Leur (1908–1942) emphasized the late date of the colonial transformation which was ultimately to end the age of cyclic history and cultural continuity, arguing that 'there is an unbroken unity in the state of Asian civilization from the seventeenth century through the eighteenth and into the nineteenth' (Van Leur 1955:284).

Even in the heyday of colonialism, there was in fact an influential opinion that the impact of the West should itself not be overestimated. 'Our influence', as Indies journalist and intellectual Willem Walraven (1992:882) memorably wrote on Java in 1941, 'does not amount to a scratch on a rock'. Orientalist colonial scholar J.H. Boeke (1884–1956), in a book on economic and social 'dualism' in Indonesia that would influence a generation of writers on underdevelopment, proposed that due to profound cultural differences, Indonesian society could never develop along Western (capitalist) lines and was destined to remain permanently poor and irrational (Boeke 1940, 1953). In so far as colonialism had made a difference, Boeke argued, it had been to reinforce, rather than erode, traditional institutions.

When, in the wake of the Second World War, decolonization came to the region more quickly than anyone had thought possible, and when the democratic systems initially created in the newly independent states began to crumble, some historians responded by thinking in terms of Southeast Asian history having 'found a way back to its own moorings' following a temporary 'diversion' into the alien domains of (colonial) bureaucracy and (post-colonial) democracy (Benda 1964:453). Western power had been nothing more than 'a thin layer resting on top of large and essentially intact societies' (Smail 1961:101), and the rapid breakdown of Westernized institutions once it was removed reflected the inevitable resurgence of old indigenous norms favouring personalistic, charismatic, and authoritarian government.

Was the colonial era, then, nothing more than 'a fleeting, passing phase' in the history of Southeast Asia, as the last British governor-general of Malaya once put it (Tarling 2001:199)? Today, in the aftermath of a whole series of further revolutionary changes in the late twentieth century – the deadly struggle between communism and capitalism in the 1950s and 60s, the astonishing and unpredicted economic miracle of the 70s, 80s, and 90s, the return of surprisingly stable democratic government in the Philippines, Indonesia, and now perhaps Burma – it is much harder to entertain any idea of Southeast Asia's history having any 'natural' direction to which its course automatically returns. Meanwhile, cumulative historical and archaeological research on the period before 1500 has undermined any idea of a long precolonial stagnation by illuminating dramatic early transformations that were once partly hidden from view: mass religious conversions that deeply affected the life of peasants as well as elites, periods of commercial expansion long before the colonial trade boom, and changes in technology and agriculture that had nothing to do with European science.[1]

Yet for all the sophistication and nuances of our emerging view of Southeast Asia's *longue durée*, it remains clear that both the geographical and the civilizational (cultural) structures which Braudel saw as 'encumbering' European history do find ready parallels there. On the geographical front, the classic tripartite agricultural ecology of Southeast Asia, based on wet rice in irrigated fields, dry rice on rotational swiddens, and commercial arboriculture, has been a constant from earliest times almost up to the present, affecting the size and distribution of populations and the development of states and trade. The three farming systems reflect constraints imposed by the infertility of Southeast Asia's tropical forest soils, constraints which each of the three provides a different way of overcoming (Henley 2002, 2008). It is partly through growing awareness of environmental influences on Southeast Asian history that the modern historiography of the region has been increasingly inspired by the spirit of Braudel. Many authors have even discerned a direct geographical parallel between Braudel's 'Mediterranean World', in which water, rather than land, is the connecting and unifying element, and the maritime world of

1 Bellwood (1985); Boomgaard (2007b); Lieberman (2003, 2009); Reid (1988–93); Hall (2011).

Southeast Asia with its shallow, readily navigable seas, its extended coastlines, and its island chains (Sutherland 2003).

Some of the environmental conditions which shape and limit human activity are affected by natural changes which, although cyclic in nature, take place over timespans that place them in the realm of the *longue durée* rather than that of Braudel's 'history of cycles'. Long-term climate change is the prime example. Evidence from growth rings in teak and other deciduous trees from Java, Burma, Thailand and Vietnam reveals extended periods of repeated drought and climatic instability which have been linked with the collapse of Angkor in the fourteenth century (Buckley et al. 2010), with the 'seventeenth-century crisis' in Indonesia (Reid 1990b:654–657), and with political crises in Burma and Thailand in the mid-eighteenth century (D'Arrigo et al. 2011:4).

Turning to persistent structures in the cultural sphere, it remains an undeniable fact that religious and political models derived from South Asian sources structured the societies and states of Southeast Asia during more than a millennium of 'Indianization'. In the Theravada Buddhist countries of the Southeast Asian mainland, their influence is still strong today. In island Southeast Asia, local conversions to Islam in the fifteenth and sixteenth centuries established paths of cultural development which were to shape the lives of hundreds of millions over the subsequent half millennium (Ricklefs 2006, 2007).

Not all cultural goods that initially developed major traction in the region, of course, went on to have such lasting influence. In the mid-twentieth century, the ideology of communism unleashed cataclysmic political forces in many countries and seemed set to shape the future of the whole region; but today it already appears a spent force. By the same token, not all environmental forces shape human history only as steady constraints in the long term: some, such as the volcanic eruptions, earthquakes and tsunamis which Indonesians in particular are once more all too familiar with in the early twenty-first century, are more sudden, more intermittent, and more violently destructive in their effects than is any war or economic crisis. Others, such as the quasi-periodic El Niño climatic oscillation, occur in cycles of short duration: episodes of drought and flooding in Indonesia, we now know, are linked to the El Niño phenomenon with its average recurrence interval of about five years (Boomgaard 2007b:100–101, 122–123).

Environmental influences on human activity, then, may be much less permanent and invariable than Braudel assumed. But even when intermittent, they can still shape the history of the *longue durée* – whether because they occur at short and regular enough intervals to form systematic and predictable constraints on subsistence, as in the case of El Niño, or because their effects on all aspects of life are so destructive that they continue to have repercussions for centuries, as in the case of certain seismic and volcanic events which will be described in the following chapters.

This book aims to underline the importance of persistent patterns and structures in Southeast Asian history, the effects of which are felt in multiple domains of human activity and at periods widely separated in time, including the present. These patterns and structures can help us to understand both the course of the region's history, and what makes it distinctive in relation to other parts of the world. Our book attempts to identify and explore some of them, yet without falling into the trap of essentializing them – that is, of suggesting that they are permanent and inescapable, or that they are unique and intrinsic to particular peoples or cultures.

Examples of persistent and characteristically Southeast Asian historical patterns explored by our contributors include environmentally determined features such as the periodic natural disasters associated with the region's tectonic structure (Chapters 4 and 5), and the structural openness to seaborne commerce which is dictated by its geography (Chapters 7, 8, and 9). Persistent economic, political and cultural features considered include the role played over many centuries by ethnic Chinese traders as an occupationally specialized business minority in the region (Chapter 10), and the resilient pattern of social organization known as patronage or clientelism, whereby political power is based on an unequal exchange, between individuals of higher and lower status, of resources and loyalty on the one hand for protection and security on the other (Chapter 11). Other chapters (2, 3, 6 and 12) focus less on persistence and continuity than on long-term change, respectively in the areas of vegetation cover, domesticated animal populations, human fertility, and visual representations of Indonesia.

Following our introduction, a series of chapters focus on the environmental and demographic themes central to much of Boomgaard's work. Greg Bankoff's 'Deep Forestry: Shaping the *Longue Durée* of the Forest in the Philippines' (Chapter 2) showcases a 'biocentric' approach to environmental history in which nature, not man, occupies centre stage, and human activity figures as just one of many factors influencing the development of Philippine forests. While the sheer scale of human forest clearance has overshadowed other changes in the period since 1946, in the long sweep of forest history this is a mere moment. Nor does the modern era of anthropogenic deforestation mark the end of a primordial equilibrium. At earlier periods natural disturbances, such as climatic cycles and changing fire regimes, already made the forests of Southeast Asia dynamic in extent and composition. When humans appeared on the scene, moreover, their influence interacted dynamically with that of natural agents such as fire, hydrology, and termites. Bankoff concludes by stressing the roles which forests still play in 'mitigating the potentially destructive power of natural forces in an increasingly less predictable world'.

Chapter 3, 'Breeding and Power in Southeast Asia: Horses, Mules and Donkeys in the *Longue Durée*', by William Clarence-Smith, continues the biocentric theme in a piece which, although styled as a tribute to pioneering work in the same area by Boomgaard (2004, 2007a), introduces two quite radically new propositions regarding the origins of Southeast Asia's horse and donkey populations. The first proposition is that Tibet, not India or China, has been the main external source of equine diffusion into the Southeast Asian region. The second is that historians have tended to overestimate the significance of equids brought in from outside the region, both in terms of the numbers imported and in terms of their genetic contribution. Where successful new breeds appeared in Southeast Asia, they were mainly the result of selection and management, not of crossbreeding with imported stock.

Chapters 4 and 5 deal with a class of natural events that are intermittent and catastrophic, yet may ironically may be reckoned among the most persistent and characteristic influences on the history of Southeast Asia in the *longue durée*. The Indonesian archipelago is to a large extent structured by a 'ring of fire', an arc of volcanoes running from Sumatra via Java and the Lesser Sunda Islands toward the Philippines. Well-known volcanic disasters include the eruptions of Krakatau in 1883 and Tambora in 1815, which affected not only Southeast Asia but also the rest of the world. The Tambora eruption launched so much ash into the air that temperatures were lowered worldwide, so that 1816 was called 'the year without summer' (De Jong Boers 1995). The movements of colliding continental plates beneath the ring of fire make the region tectonically active, causing earthquakes and tsunamis. Together, volcanic eruptions, earthquakes and tsunamis have produced historical ruptures, fundamental changes with long-lasting effects.

In Chapter 4, 'Under the Volcano: Stabilizing the Early Javanese State in an Unstable Environment', Jan Wisseman Christie investigates the impact of volcanic eruptions on political relationships and religious attitudes in Central Java during the tenth century. Basing herself on a meticulous analysis of contemporary inscriptions, she concludes that a series of volcanic eruptions at the beginning of the tenth century destabilized the kingdom of Mataram with two lasting consequences. One was that the kingdom abandoned Central Java and moved its centre to East Java, where a new and long history of state formation was initiated. A second consequence, Wisseman Christie argues, was a lasting change in religious attitudes. Mahayana Buddhism and its central place of worship, the Borobudur complex, were abandoned and replaced by the worship of the Hindu god Shiva, together with that of Javanese mountains and ancestors. Buddhism, it seems, had no answer to the volcanic violence that overtook Mataram. Instead the god of destruction, Shiva, won in prestige, as did local ancestor-deities associated with mountains. In the centuries that followed, it was Shivaism that dominated the rituals sustaining the continuity of royal rule in East Java.

The tsunami disaster that struck northern Sumatra on 26 December 2004 stimulated historians to search for evidence of similar events in earlier times. In Chapter 5, 'History and Seismology in the Ring of Fire: Punctuating the Indonesian Past', Anthony Reid argues on the basis of recently discovered geomorphological evidence that northern Sumatra was very likely hit by a devastating tsunami around the beginning of the fifteenth century. As a result Buddhism was literally swept away, leaving hardly any trace of its previous existence, while the spread of Islam accelerated. In Central Java, a hitherto forgotten tsunami of 1618 may likewise have been the cause of major social and political change. Reid suggests that the well-known goddess of the southern sea, Niai Loro Kidul, with whom the Central Javanese kings maintained a mystical relationship, should be associated with the threat of tsunamis, to which people living in coastal plains were particularly vulnerable. The chapter concludes by noting that the recurring damage caused by eruptions, earthquakes and tsunamis offers a new part of the solution to the old puzzle of low population growth in precolonial Indonesia.

Chapter 6 is Linda Newson's 'The *Longue Durée* in Filipino Demographic History: The Role of Fertility prior to 1800'. Drawing on extensive research in published and archival sources, Newson follows the example set by Boomgaard in the Indonesian context by arguing that in the Philippines too, variations in population density and growth over time and space can be ascribed partly to variations in the frequency of traditional fertility control practices such as induced abortion and infanticide. Since the evidence of indigenous birth control is strongest in the oldest available sources, it is clear that essential facts about the dynamics of demographic change in the Philippines do not become apparent unless a very long-term historical view is taken.

Chapter 7, 'Glimpsing Southeast Asian *Naturalia* in Global Trade, c. 300 BCE–1600 CE', by Raquel Reyes, forms a natural bridge between the environmental themes of earlier chapters and the economic, social and political themes of those which follow it. Reyes explores the little-known history of the trade in exotic natural objects – from bezoar stones to rhinoceros horns – between Southeast Asia and the rest of the world. Despite its obscurity, this trade is an ancient one and true to the panoptic spirit of Boomgaard's own work, Reyes surveys it over a timespan of fully nineteen centuries. She also takes in both the European and the Chinese markets for Southeast Asian *naturalia*, and considers the impact of the trade on the development of Western science, on Chinese medicine, and on the environments and societies of Southeast Asia.

In Chapter 8, 'Ages of Commerce in Southeast Asian history', David Henley provides a brief critical survey of the growing literature on very long-term Southeast Asian commercial and political cycles inspired by Anthony Reid's seminal *Southeast Asia in the Age of Commerce* (1988–93). Two responses to Reid receive particular attention: that of Geoff Wade (2009a), who identifies an earlier period of trade-based economic and political florescence (900–1300) to place alongside Reid's early modern Age of Commerce (1450–1680); and that of Victor Lieberman (2003, 2009), who argues that the long-term development of mainland Southeast Asia followed a very different pattern from that of island Southeast Asia, on which Reid's argument is mostly based. The chapter concludes by suggesting that the idea of environmental change as an underlying pacemaker of economic and political cycles may help to reconcile the competing maritime and mainland perspectives.

Heather Sutherland provides further critical reflection on the Reid paradigm in Chapter 9, 'Pursuing the Invisible: Makassar, City and Systems'. Here Sutherland uses detailed historical evidence from Makassar to evaluate and nuance Reid's proposition that a comprehensive economic and demographic crisis, marking the end of his early modern Age of Commerce, followed the establishment of VOC control over Indonesia's international trade in the mid-seventeenth century. She finds that although the conquest of Makassar by the Dutch in 1669 did have serious economic consequences, it did not by any means leave the city a shrunken, inward-looking shadow of what it had been in its spice-trading heyday. The economy of its former political hinterland, meanwhile, was little affected by Makassar's fall, since regional commerce simply shifted in large part to different harbours. Continuity, more than change, characterizes South Sulawesi's history in the seventeenth and eighteenth centuries.

Chapter 10, by Kwee Hui Kian, continues and concludes the commercial theme with a condensed history of the development of ethnic Chinese trading networks in Southeast Asia from 1400 to 1850. Successive waves of Chinese sojourners and settlers in the region engaged in long-distance trade with China itself, and gradually also penetrated into regional trade and local markets, before in some cases ultimately becoming directly involved in the production of key trade commodities: sugar, gambir, pepper, rice, tin. Their commercial networks were knitted together partly by intensive cooperation among migrants sharing the same surname or geographical origin, which enabled newcomers to obtain jobs and cheap credit. Equally important to their success, however, was the political context that enabled Chinese migrants to fill vital niches in the Southeast Asian economy. Power-holders, first indigenous and later European, allowed and even encouraged Chinese migrants to occupy strategic commercial positions. Kwee's long-term perspective reveals how the basis was laid for the dominant role Chinese entrepreneurs and traders were to play under colonial and postcolonial capitalism.

In Chapter 11, 'From Contest State to Patronage Democracy: The *Longue Durée* of Clientelism in Indonesia', Henk Schulte Nordholt reviews the history of patron-client relationships as a structuring element in political life over two millennia. Although his focus is on Indonesia, his observations are largely generalizable to the rest of Southeast Asia. Having argued elsewhere for a similar continuity with respect to forms of political violence (H. Schulte Nordholt 2002), here he proposes that the phenomenon of patronage forms a continuous link between the precolonial contest state and today's patronage democracy. The leaders of modern civil militias, he believes, resemble in many ways the 'men of prowess' who dominated traditional political systems. Nevertheless the form of the patron-client relationship has changed over time, adapting itself to new institutional conditions. Whereas in precolonial times relationships between leaders and followers were unstable, in the colonial period they were to a large extent stabilized by incorporation into a system of indirect rule. With national independence, clientelism did not disappear. Indeed, it can be argued that new chains of patron-client ties, stretching from the capital into the villages, actually glued the young and fragile nation-state together. Similar relationships were the oil that kept the New Order running, despite its image as a strong bureaucratic state. Reviewing recent developments, Schulte Nordholt concludes that patronage will probably continue to structure the political system in the foreseeable future.

The twelfth and final chapter, by Jean Gelman Taylor, concentrates on the visual representation of colonial encounters over a period of three centuries. Inspired by Peter Burke's remark that paintings are 'painted opinions', Gelman Taylor traces the visual experiences of the colonial encounter through paintings, photographs and film. Not all of the images referred to are reproduced in this book; to see the remainder, and many more, the reader is invited to visit (virtually or otherwise) the relevant collections of the KITLV, KIT, and Rijksmuseum. For discussion purposes, Gelman Taylor divides her materials according to six thematic series. The first depicts ships, harbours, markets, and traders, grouped under the rubric 'From ship's deck to shore'. The second depicts encounters with Asian arts. The third contains portraits, the fourth, landscapes, and the fifth, depictions of historical events. The final series consists of photography and offers 'glimpses of modernity'. Gelman Taylor argues that photographs, on a greater scale and with greater social impact than portrait paintings before them, 'document the emergence of the named individual in Indonesian life', and that the camera 'is an indelible part of contemporary Indonesian culture'.

By following Peter Boomgaard in taking a *longue durée* perspective on Southeast Asian history, it is possible to overcome the compartmentalization of conventional historical periodization, which often obscures underlying structures. Each historical period tends to have particular characteristics ascribed to it, together with a particular analytical vocabulary that is 'owned' by an authoritative group of experts. With regards to the Indonesian archipelago, the following 'compartments' can be identified. Early history was traditionally owned by archaeologists, who developed their own discourse about forms of kingship and processes of Indianization. The early modern period was for a long time dominated by VOC historians, while the nineteenth century was the domain of experts on agrarian exploitation and colonial expansion. The late colonial period was the hunting ground of historians who traced the rise of nationalism.

In the middle of the twentieth century, shorter periods, each with its own contingent of experts, succeeded one other more rapidly: the Japanese occupation, the national revolution, and the 1950s, a decade typically discussed with an emphasis on nation building. Writers on the 1970s and 1980s, for their part, tended to focus on the political economy of the strong state. It was possible to write on nation building in the 1950s without considering the development of nationalism in previous periods, and to concentrate on the revolution without bothering much about its effects on the events of later years. In this way the lives of Indonesians who were born in the 1920s, and lived through the various periods, were cut into disconnected pieces. Moreover, the connection between similar relationships (such as clientelism) in different eras was often overlooked. Against such compartmentalization, a *longue durée* perspective opens up the possibility of identifying patterns and changes which transcend the confines not only of particular decades, but of the nation-state itself.

A great many topics have not been addressed in this volume, which of course cannot pretend to represent the whole state of the art in its very broadly defined field. In the future, new studies of long-term continuity and change in literary themes, religious practices, eating habits, and many other aspects of life, as well as new comprehensive *longue durée* studies of particular places and regions, will no doubt uncover further hidden layers in the history of Southeast Asia. But as far as the present state of the art is concerned, our book does at least give a taste of some of the latest ideas and insights. As such, it is an invitation to undertake further exploration of Southeast Asian history in terms of its *longue durée*.

CHAPTER 2

Deep Forestry

Shaping the Longue Durée *of the Forest in the Philippines*[1]

Greg Bankoff

The forest is a dynamic place, a complex ecosystem that has adapted over the *longue durée* to climatic and edaphic conditions. More than trees, the forest is also about plants, animals and the other agents that live in, use and consume the forest. Yet when it comes to writing forest history, the complexity of this system mainly vanishes: forest history becomes largely a narrative of human destruction of the forest over time. But there are other shapers of the forest. Climate and soil, to begin with, determine where and what types of species grow or predominate. Fire, while mainly anthropogenic (human-induced), is also autogenic (naturally induced) even in tropical forest. And then there are the living agents, starting with the white ants that prune away trees that are weak and consume the dead ones. They, too, have a close relationship with human behaviour in ways that decide which species are felled and for what purpose. Finally there is humanity, with all its changing needs and wants as well as its increasing desire for land to cultivate, largely at the expense of the forest. Over the centuries, these processes have shaped the forests of the Philippines in many and varied ways.

Finding the right balance between nature and humanity in the forest has not always been an easy matter to determine. On the one hand, ecologists view people as just one among a variety of factors that alter the environment (Christensen 1989:116). Historians, on the other hand, see the forest as largely 'synthetic' (R. Nash 1972:336). If the subsequent development of environmental history as a recognized sub-discipline has been partly driven by a desire to put the science back into history (Worster 1994:ix–xi), the emergence of historical ecology has been more of an attempt to examine 'history from the viewpoint of nature' (Whitney 1994:4). The result has been a profusion of new forest histories that do justice to both disciplines.

'Deep forestry' is an attempt to put the nature back into the study of Southeast Asian forest history and combines both a bio-centric (ecology-centred) and anthropocentric (human-centred) focus. It invokes a qualified comparison with the norms and values that underlie deep ecology. As defined by Arne Naess,

1 An extended version of this chapter first appeared in *Environmental History* 18, no. 3 (2013): 523–556.

deep ecology includes among other things an awareness of the internal interrelatedness of ecosystem (Naess 1973:95–98). In particular, Aldo Leopold's exhortation (1979) to reject the primacy of human needs and instead 'think like a mountain' captured the new spirit of the time. Harking back to Leopold's injunction, deep forestry is an attempt 'to think more like a forest'. As historiography, it extends the web of relationships to include nonhuman factors. Its temporal reach, therefore, spans millennia rather than centuries. Exploring forest history from this perspective adds to both the knowledge of how the forest changed over time and the extent to which human actions contributed to those processes. Applying a deep forestry approach shows how the forests of the Philippines have been shaped as much by non-human factors as by human hands, though the balance has certainly shifted over time with human influence steadily growing. Climate, soil, fire and animals have left an imprint on the forest both through their own actions as well as through their influences on human agency.

Climate and Soils

The climate of the archipelago is controlled first of all by latitude and then by altitude. Lying between the equator and the Tropic of Cancer, the lowlands of the Philippines experience average temperatures in the mid to high twenties centigrade (Wernstedt and Spencer 1967:41). Rainfall is the most important climatic element as regards the characteristics of forest growth. Monsoonal or rain-bearing winds govern precipitation rates over the entire western side of the archipelago, the climate being marked by very distinct wet and dry seasons. Over the eastern half of the islands, rainfall is distributed throughout all the months of the year and there are no pronounced wet and dry seasons. These two different climatic regions roughly cut the Philippines in two (Wernstedt and Spencer 1967:58–62).

While there is little difference in the total precipitation between west and east, there is great variety in seasonal rainfall (Wernstedt and Spencer 1967:53–57). In general, species requiring a large amount of moisture in the soil flourish in the eastern region. The best development of dipterocarps, therefore, occurs in the eastern and northern part of Negros and in other parts of the Visayas, on Mindanao, and along the east coast of Palawan. Species that flourish in fairly dry situations are to be found across the archipelago as there are areas of relatively low rainfall in both east and west. Moreover, at elevations over 900 metres, rain falls throughout the year making the climate suitable to trees demanding high atmospheric humidity. As a rule, though, certain species prosper in one or

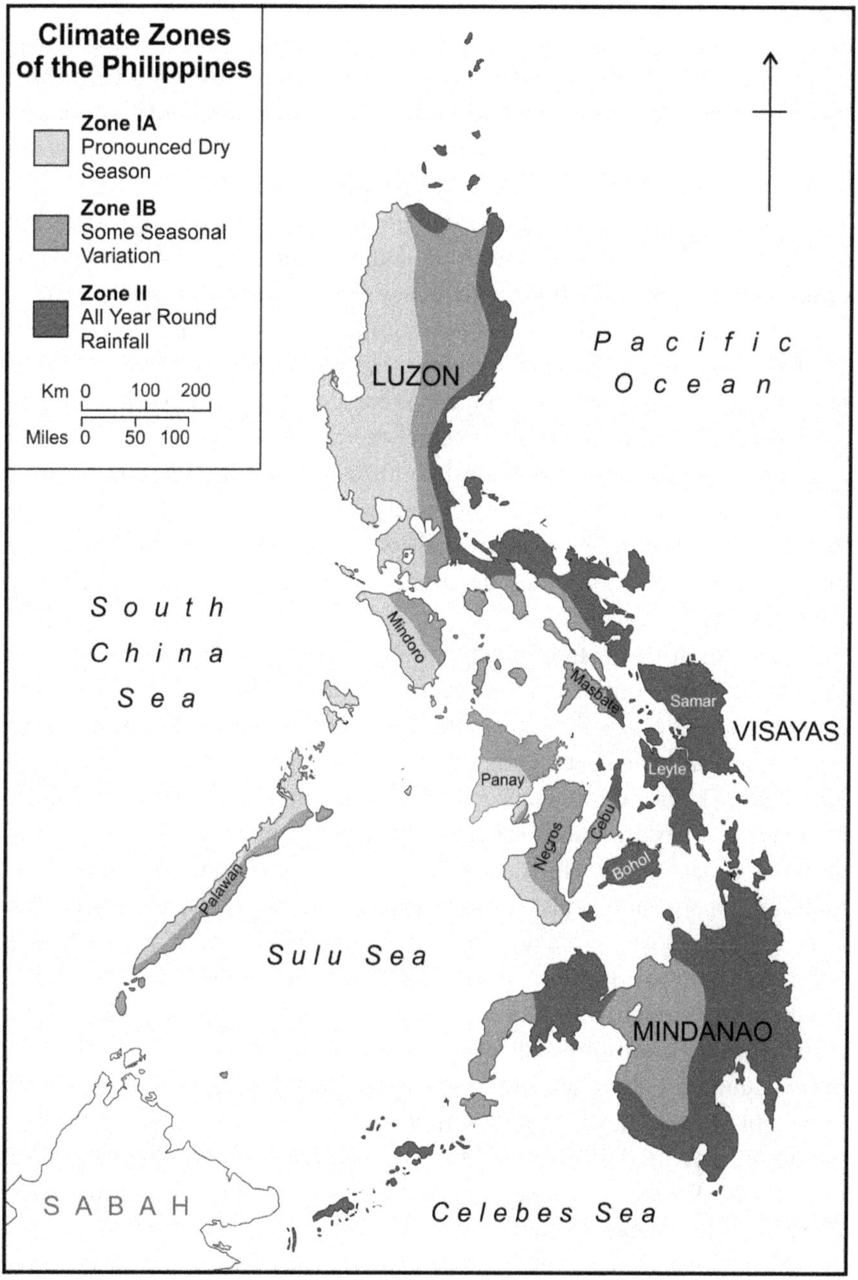
Climate Zones of the Philippines
Zone IA
Pronounced Dry Season
Zone IB
Some Seasonal Variation
Zone II
All Year Round Rainfall
Km 0 100 200
Miles 0 50 100
Pacific Ocean
LUZON
South China Sea
Mindoro
Masbate
Samar
VISAYAS
Leyte
Panay
Negros
Cebu
Bohol
Palawan
Sulu Sea
MINDANAO
SABAH
Celebes Sea

FIGURE 2.1

other region: lauan (*Shorea negrosensis*), apitong (*Dipterocarpus grandiflorus*) and guijo (*Shorea guiso*) develop best in the east, while yacal (*Shorea laevis*), narra (*Pterocarpus indicus*) and molave (*Vitex parviflora* Juss.) grow best where there is a dry period. Grasslands are also more common in areas with a prolonged dry season. The principal effect of the wind upon vegetation depends on whether it is moisture-bearing or not (Matthews 1914:3–5).

Soil and subsoil are also important to the physical characteristics of the forest and to the distribution of species. Soil affords anchorage for trees and also constitutes the reservoir from which they derive water, nitrogen and other minerals. Tropical broad-leaved trees require tremendous amounts of water and hence forests are confined to regions with adequate rainfall. Just as important as water is soil depth; in general 1.5 meters is a sufficient depth for almost any tree. If the soil is deep and moisture-retentive, species that require high moisture may be able to exist in localities of relatively low rainfall. Conversely, trees that usually require less soil moisture may be unable to exist in regions of high rainfall if the soil is sandy or shallow. All species of tropical trees grow better in soil which is minimally deep, porous, moist, warm, and rich in humus (Matthews 1914:9–14).

As a result of the distinct precipitation shadow between the western and the eastern halves of the archipelago and differences in soil depth and moisture, the forests of the Philippines are quite varied in their composition. Six types of forests have been commonly identified. *Dipterocarpaceae* are generally large trees, reaching heights up to 50 metres and with diameters of 100–150 centimetres. They are mainly evergreens and found on nearly all terrains up to altitudes of 800 meters. The molave forest, on the other hand, is composed of non-dipterocarp hardwoods such as molave and narra. Such species form more open landscapes where the dominant trees are less abundant and set further apart, are short-boled and irregular in form, and have wide-spreading crowns. Their timber is much sought after because of its strength and durability. Other forest types are more localised: mangrove on the mud flats at the mouth of rivers and along the shoreline, beach forests in coastal areas, pine forests in the upland plateaus of northern and central Luzon, and mossy-forests in the high mountain regions (Whitford 1911:17–32). However, it is not altogether clear whether the forests have always existed in quite such proportions and other forest types might have been more plentiful in the past (Whitford 1911:26–31).

Gifford Pinchot, who visited the island in 1902, provides a valuable record of how the forest looked before large-scale logging. He noted how the forest came right down to the water's edge, even overhanging the high-water mark.[2] The

---

2 LC, Pinchot Papers, box 640, 'XII Philippines':12, 16, 26, 25.

trees were of an extraordinary size, many with diameters in excess of three feet but he was too unfamiliar with the species to get any idea of their distribution.[3] There was, however, a general pattern to its appearance: Trees were shorter, smaller in diameter and 'less valuable' at the water's edge than they were

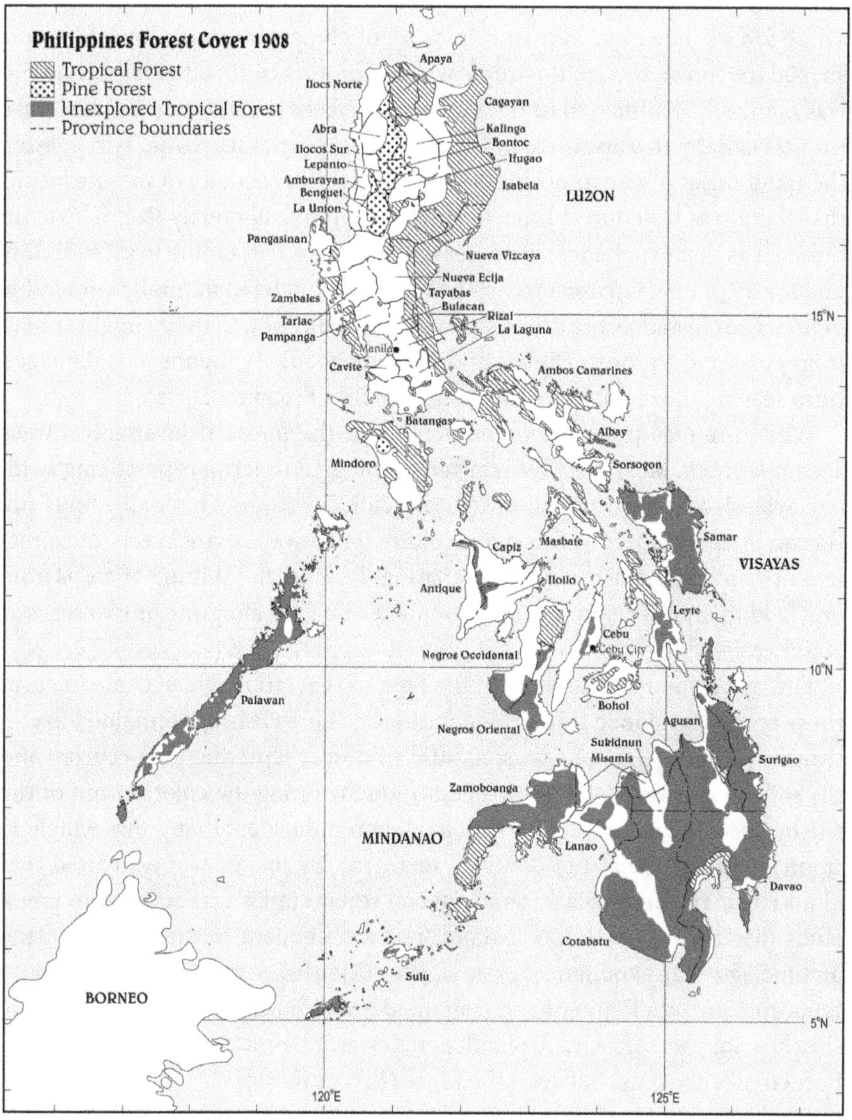

FIGURE 2.2

3 LC, Pinchot Papers, box 640, 'Philippine Islands'.

farther inland. The finest timber with the best logs was always to be found on the steepest slopes but the forest began to diminish again the higher one climbed.[4]

Fire and the Forest

These forests, however, were not the work of climate and soil alone. Fire, too, shaped the forest, though the tropical vegetation was difficult to ignite and fire was not easily maintained in the moist, shaded environment under the dense tropical canopy. In those areas of the archipelago without a distinct dry season, the rapid onset of decomposition and the relentless activity of insects meant that there was little forest litter to sustain even a deliberately lit fire. Even in those areas that experienced a dry season, it was difficult for fire to spread. Only under exceptional circumstances was the forest rendered naturally vulnerable to fire. If, for instance, the dry season was unusually hot, then fire might spread from grassland to forest (Fuller and Murphy 2006). Typhoons also damaged trees, leaving the forest more susceptible to fire (Fischer 1929:35).

While fire may precede human activity in the forest, it invariably always accompanies it. Most fires were started by indigenous farmers practicing swidden or slash-and-burn agriculture (locally called *kaiñgin*). This 'indigenous' fire was an integral aspect of forest agriculture and always shared certain characteristics such as the use of fire in preparing the land, the shifting of crops from one field to another, and the abandonment of a field after one or two seasons (Conklin 1957).

Fire had a good deal to do with the type of vegetation which came to cover those areas abandoned by swidden farmers. Where clearings remained in cultivation for more than one season, little volunteer reproduction occurred and the soil became dry and baked, a condition favouring the colonization of the site by cogon (*Imperata cylindrica*), a tall perennial grass that grows widely in tropical regions (*Report of the Chief* 1906:278). By 1914, it was estimated that almost half the archipelago, some 48,000 square miles was covered in grasslands (Matthews 1914:66). Grassland fires are frequent in those areas of the archipelago that experience a dry season (Matthews 1914:67). In the mountains, too, fire was frequent and fashioned a very particular vegetative regime (Fischer 1927:22, 1928:21). Upland peoples were pastoralists who regularly burned over the grass between the pines (Bryant 1907:9).

By the early nineteenth century, three distinct types of landscapes predominated over most of the Philippines: cultivated areas, open or semi-open

4 LC, Pinchot papers, box 640, 1902: file 9: 'Lecture on forests . . . second half'.

grasslands, and forest. The relative proportion of these areas varied according to physical features, soil quality, and the density of human population (Matthews 1914:67). As a general rule, agricultural lands occupied the fertile river valleys and adjacent slopes, grasslands spread over the rocky or steep slopes, and forests covered the mountains. While some fires were due to carelessness, most were deliberately lit by hunters eager to attract game (*Report of the Chief* 1906:278; Matthews 1914:4). The intense heat of these fires killed off surviving stumps as well as any seedlings, causing the tree line to slowly recede over time (Matthews 1914:67).

Europeans changed fire regimes, both intentionally and unintentionally. On the one hand, fire was the inadvertent consequence of the new steam-powered machinery used in commercial logging at the beginning of the twentieth century (Fischer 1923:38). Fires were more frequent during the dry season and especially in areas that had been logged (*Forest District* 1905:27). On the other hand, fires might be intentionally set to clear the ground before logs were skidded to a landing. Uncontrolled fire in logged over areas destroyed all forest growth. George Ahern (1913:11), first chief of the Philippine Island Bureau of Forestry, noted how: 'Even under most favourable conditions following logging operations the young seedlings have difficulty in withstanding the strong isolation to which they are exposed, but after such a destructive fire conditions are rendered altogether hopeless'. Fires in logged-over sites were frequent events by 1930, and they were apparently 'becoming more and more of a problem'. Sometimes, too, fires were deliberately lit by forestry officials 'to clear up old caiñgins' or as a preventative measure before the start of the fire season (*Forest District* 1905:31; Fischer 1925:15). Europeans also shaped the forest: they abetted the spread of fire to locales where it had rarely been seen before, and, by clear cutting burnt over areas, encouraged the spread of grasslands.

Anay: Shepherds of the Trees

The termite or 'white ant' is integral to the very fabric of the forest. Known generically as anay in the Philippines, more than 1,500 different species are distributed throughout the Tropics. In tropical forests, anay remove ground litter and enrich the soil by working leaf-matter deeper into it. Most species only attack dead plant material but a few are serious pests to living trees (Dammerman 1929:26–27). In the Philippines, there are 54 species of termites. The most common species are the mound-building termite (*Macrotermes gilvus*) that constructs extensive systems of covered passages to attack dead wood including structural woodwork. The most destructive, however, is the

Philippine milk termite (*Coptotermes vastator*). Other common species include the Los Baños termite (*Microcerotermes losbanosenses*) that attacks waste lumber and woodwork such as roof joists and rafters, and the Luzon point-headed termite (*Nasutitermus luzonicus*) that prefers to construct nests on tree trunks. Altogether these and other species are widely distributed, cause significant structural damage to buildings, and are considered serious pests (Acda 2004).

In the past, anay were the real masters of the shade beneath the forest canopy: They shaped the distribution of species growing in forests by influencing which trees people were willing to harvest. Of the hundreds of tree species to be found in the Philippine forest, only 30 to 40 were regularly felled. Over hundreds of years, the effect of felling only a few tree species was the 'complete exhaustion in certain provinces of the best kind of timber'.[5] Trees that anay did not eat were the species mainly used by indigenous people.[6] After the Spanish colonized the islands in 1565, white ants continued to influence which tree species were logged. There was an increasing demand for hardwoods resistant to anay for use in the construction of new urban centres and upland fortifications. As Gifford Pinchot noted in 1903:

> You find the ants everywhere, in the towns and out of them, and you find also that they have controlled the whole process of lumbering in the Islands until now, and the reason why the cut-over land is in its present condition in the Philippines, the reason why certain species are getting scarce in certain places, are almost entirely due to the fact that their timbers resist the white ant. What does not resist the white ant has been let alone.[7]

Similar sentiments had been expressed 26 years earlier by Spanish officials who were aware of the situation and appreciated that 'in Tayabas and other localities of this archipelago woods such as mangachapuy, betis, banaba and others of the most sought after species ... are already very scarce'.[8]

Accessible stands of such timber were increasingly hard to find and mature, seed-bearing trees became exceedingly rare. This culling of the largest trees led to a form of genetic erosion, diminishing the contribution of a particular

5 LC, Pinchot papers, box 640, 1902: file 9: 'Lecture on forests; box 640, 1902: file 10: 'Lecture on forests ... second half', 'Lecture on forests ... first half'.

6 LC, Pinchot papers, box 640, 1902: file 10: 'Lecture on forests ... first half'.

7 LC, Pinchot papers, box 640, 1902: file 9: 'Lecture on forests; box 640, 1902: file 10: 'Lecture on forests ... second half', 'Lecture on forests ... first half'.

8 PNA, Ministerio de Ultramar ... Islands, 7 August 1877.

species to the general biomass, and to the attenuation of remaining stands with consequences for the fauna and flora dependent upon them. Pressure on resistant tree species only diminished after 1910 when extensive field trials at the newly established timber testing laboratory at Los Baños discovered that creosote was an effective treatment against white ant. Subsequently, loggers harvested a larger number of timber species for construction purposes.[9]

In addition to white ants, other animals have also shaped forest cover. Cattle, sheep and horses have not so much led to forest destruction as hindered its regrowth. Domestic cattle were never numerous in the Philippines until the late nineteenth century and sheep were confined to upland areas (Edgerton 1982:375). The horse, too, was not native to the Philippines. First brought from New Spain and subsequently imported in much greater numbers from China and Japan, horses were successfully bred on 'great stock farms' in the seventeenth and eighteenth centuries (Bankoff 2001). Unlike cattle, horses played a significant role in the economic life of the wider society and horse-ownership among indigenous peoples was widespread by 1800. For most of the year, however, these horses were moved to communal pastures in the uplands where they were left to scavenge for whatever they could find to eat. In this way, the horse was an unwitting agent hindering forest regeneration and turning cleared or logged areas into grasslands.

State and Forest

Making a division between non-human and human agents is somewhat arbitrary in deep forestry: It is often the symbiotic relationship between the non-human and human that proved to have such an important impact on the shape of the forest. Human actions in the Philippines undoubtedly had unforeseen results on the forest when coupled with the forces of the natural world, extending the reach of fire, bestowing on white ant the power to determine which species were cut or, as in the case of horses, hindering regrowth. All of which changed the composition of the forest beyond recognition. However, this human impact evolved in relation to the wider historical context, as a result of changes in demand for timber products and, ultimately, as the composition of the forest itself changed over time. In particular, the formation of the state in the mid-sixteenth century and the commercialization of forestry beginning in the mid- to late-nineteenth century were decisive turning points in the forest history of the Philippines.

9 NARA, General records of the Bureau of Insular Affairs, box 768, file 11896.

Not all timber has the same qualities and some are more conducive to human purposes than others. Certain species, therefore, have been felled more often than others. Trees or timber products were used among other purposes for housing, transportation, furniture, utensils, writing, medicine, as a source of heat and even as clothing. The tools of agriculture were primarily wooden as were largely the weapons of war. All this wood initially came from the forest so that the history of the forest is largely commensurate with the modern history of the societies that lived in and about it.

Little is known about the precise impact of pre-colonial societies on the environment. While no hydraulic states emerged in the archipelago to rival those found elsewhere in Southeast Asia, Metal Age chiefdoms existed and were able to expand quite rapidly with the advent of a foreign trade in Chinese porcelain after the tenth century. The added wealth this trade provided even created the conditions whereby several larger scale, inter-regional polities were able to develop at Manila, Cebu, Sulu, and Magindanao. Chiefly-power rested on engaging in alliance-building exchanges through attracting foreign luxury goods and augmenting the available labour supply by intensive maritime raiding. Raiding, in turn, prompted the increased construction of coastal fortifications, an expansion of metal weapon production, the adoption of foreign military technology, and the emergence of a specialized warrior class – all of which required wood (Junker 1999).

Warfare was a highly developed aspect of daily life in the pre-Hispanic Philippines and most of these weapons were made, at least in part, from wood (W. Scott 1994:147–151). Many Filipinos spent much of their time on water, and many military engagements also took place at sea. In these circumstances, the construction of warships was a highly developed aspect of military technology. These karakoras were sleek, double-ended vessels. They mounted a square sail and had double-outriggers on which paddlers provided speed in battle. They were light, flexible, extremely manoeuvrable and perfectly suited to the maritime conditions in which they operated. They were also fighting machine par excellence as Europeans later learnt to their discomfort. Karakoras reached bursts of speed of between twelve to fifteen knots in contrast to the five or six knots that a European galleon made (W. Scott 1981:5–10). As one Spanish observer ruefully noted in 1667, 'Their ships sail like birds, while ours are like lead in comparison' (Combes 1897:70–71).

The preferred wood used in the construction of karakoras was molave because of its strength. The prevalence of these crafts and the amount of wood felled for their construction is difficult to gauge. However, the Maguindanaos of Mindanao were able to muster a war fleet of a hundred such vessels to raid the Visayas in 1602, and Rajah Bongsu of Jolo apparently set out for southern

Luzon in 1627 with a fleet that carried 2,000 fighting men (Corpuz 1989:141–143). However, despite the fact that warfare was endemic in these pre- or proto-state societies, the environmental effect on the forest was likely to have been localized and limited given the weaponry available and the low population density (Junker 1999:62).

The first large-scale, permanent revolution in land-use is usually associated with the development of a core region and the emergence of a state system in the Philippines after 1565. On the one hand, this is a narrative about the construction of urban and municipal centres as sites of administration or evangelization. On the other hand, it is about the development of an early agricultural market and the introduction of new crops from the Americas. The erection and maintenance of this state so far from Europe also required defence. The subsequent construction of ships and forts intensified the demand for timber. However, simply equating state formation with deforestation is too crude an analysis. Not all wood serves human purposes equally well at any given time: particular woods were selected for specific purposes and felled regularly, while others were viewed as largely worthless and left largely alone (Boomgaard 1988:58–87).

In the Philippines, the cities and towns that are so characteristic of Spanish colonialism the world over were mainly built from woods. Even the capital, Manila, founded in 1571, was initially a collection of wooden structures (Morga 1903–1909:136–143). The indigenous house, the bahay-kubo, was primarily constructed of bamboo (*Bambusa blumeana* Schultes. f.), roofed with nipa palm (*Nipa fruticans* Wurmb.), and lashed together with rattan (*Calamus maximus* Blanco) (Alarcón 1991:23–77). They were raised on hardwood poles known as haligues, generally made of molave because of their termite-resistant qualities (W. Scott 1994:57–62). Spaniards continued to use the same materials, adapting them to the architectural forms of churches and other public buildings (Gentil 1903–1909:207).

Defence, too, required a constant source of newly-cut timber. The colony needed to defend against Dutch fleets and Moro (Muslim) raiders as the Spanish presence dragged the archipelago into a wider arena of conflict and religious antagonism (Warren 1985). The aim of the Dutch East Indies Company (VOC) was to hinder trade by blockading Manila, intercept the bullion-loaded Acapulco galleons and thwart Spanish attempts to establish a permanent presence in the Spice Islands (Muijzenberg 2000:12–15). War in the South was aggravated by Spanish attempts to subjugate the Muslim polities. Hostilities took the form of Moro raids and Spanish retaliatory expeditions that persisted into the nineteenth century (Corpuz 1989:135–158). What the Spaniards needed to defend their new possessions in the East were ships. Mindful of the need for suitable wood, early accounts of the islands were quick to appreciate their potential for shipbuilding (Mirandola 1903–1908:225; Sande 1903–1909:59). Not only was there an 'abundance of wood for all kinds of vessels' but Filipinos proved to be 'very skillful in making ships and fragatas' (Vera 1903–1909:206). By 1616, six out of the seven galleons stationed at Manila had been built in the islands (Pineda 1903–1909:180). These vessels also continually required regular refurbishing or replacement (Pineda 1903–1909:171–173).

How much timber was consumed over the decades is difficult to estimate. Some idea of the scale of the whole enterprise can be gleaned from the labour that was levied to meet these demands. As part of the corvée requirement indigenous people rendered the colonial state, municipalities had to provide people to work in the shipyards or cut timber (Cushner 1971:117–126). So great was the amount of wood required that it caused revolts, most notably in 1614 and 1649 (Corpuz 1989:124–128). Certain species of tree near shipyards were felled and their timbers used in shipbuilding and repair. Molave and other valuable hardwoods soon became increasingly difficult to find. Already by 1621, Hernando de los Rios Coronel commented on how difficult it was to 'find the necessary timbers of the forest' (Coronel 1903–1909:203).

The threat, however, was not only external. Spanish control over some parts of the archipelago remained tenuous right into the nineteenth century. Especially in the Cordilleras of central Luzon, frontier communities evolved based on forts and military garrisons, a *presidio* (fort) society. Subject to missionary contact and irregular military expeditions since the sixteenth century, Governor-generals did not try to impose colonial order on upland peoples until the nineteenth century (W. Scott 1974:3–4). Even then, the Spanish presence remained primarily a military occupation based on garrison towns and forts, dependent on lowland sources for provisions and guns to enforce compliance (W. Scott 1974:296). Like the naval craft that defended the colony from the sea, the forts that straddled the highlands were also constructed of timber and were often elaborate affairs, palisades with parapets protected by moats, earthworks and outer stockades (W. Scott 1974:272–273). Given the number of presidios situated at strategic locations all over the Cordillera as well as in other parts of the archipelago, the amount of timber required in their construction and maintenance was considerable. Again the use of valuable hardwoods further contributed to the reproductive erosion of certain species, reducing the genetic diversity of the forest by leaving only less desirable individuals to produce seed. A report on the state of the forest in 1877 blamed the unregulated activities of woodcutters for a scarcity of 'valuable trees of large dimensions', and the preponderance of 'juvenile ones' which had led to the 'impoverishment' of the forest.[10]

10 PNA, Ministerio de Ultramar … Islands, 7 August 1877.

All this amounted to a very changed landscape in many parts of the archipelago by the late nineteenth century. The extent of deforestation is difficult to reconstruct. A detailed provincial breakdown of forest cover was published by the Spanish forestry department in 1875. According to this estimate, 70% of the islands was still forested after more than 300 years of Spanish rule. This percentage, however, is based on the total land area of the present nation state and not on the territories under effective colonial administration. Until the early twentieth century, much of the largely untouched forests of Mindanao lay beyond the effective control of authorities in Manila. Subtracting the latter's some nine and a half million hectares from the approximate national total of 30 million hectares reveals the true extent of deforestation by the end of the nineteenth century. Using this formulation, about half the forest cover in the territory under Spanish control had already disappeared prior to the American era (Vidal y Soler 1875:40; Bankoff 2007c).

The Commercialization of the Forest

What fundamentally changed the nature of the human impact on the forests of the Philippines was the commercialization of the timber trade in the second half of the nineteenth century. While the volume of timber cut during the American colonial period increased markedly, especially after 1918, three-quarters of it still went to supply the domestic market just as it had under the previous regime. Historians have long argued that the Americans introduced scientific forestry, but an official forestry service was established in the Philippines in 1863, eighteen years before the United States established its Division of Forestry (Roth 1983; Tucker 1988). Forest regulations in the archipelago long preceded the establishment of a forestry service. The *Recopilación de las Leyes de Indias* had two objectives: protecting customary access to the forest and encouraging its conservation. Spaniards in the nineteenth century regarded the spirit of this legal corpus and its regard for indigenous rights as constituting 'the glory' of their dominion in the archipelago (Vidal y Soler 1875:59–60, 81). Unfortunately, such laws were largely disregarded. Moreover, most new laws enacted during the first half of the nineteenth century were to encourage agricultural expansion usually at the expense of the forest (Vidal y Soler 1875:65, 67). It was the flagrant disregard for forest laws, the sheer scale of the destruction, the shortage of timber in many areas, and public criticism that persuaded the Spanish colonial administration to establish a forestry service in 1863.[11]

11 Vidal y Soler (1874:20); Jordana y Morera (1891:228); LC, Pinchot Papers, box 586, 'Regino Garcia':9.

Other forces were at work in the forests by then, too, not least the emergence of a commercial market for wood and the perceived need to manage the remaining stands in a scientific manner. Certainly a timber market had emerged in the archipelago by mid-nineteenth century (Burzynski 2002:171–180). The stimulus for this market was the growth of Manila and to a lesser extent other urban centres (Foreman 1899:355). The rapid urbanization of the capital generated a strong demand for timber that soared in the aftermath of major disasters such as earthquakes (1863 and 1880) and fire (1870). Owners with substantial houses to repair needed timber; prices rose and local merchants seized the opportunity to make substantial profits (Bankoff 2012).

Along with the growing demand for timber and the increasing problems of supply was the realization that the forests of the Philippines had to be managed more systematically. The model that Spanish foresters had in mind was a programme of artificial reforestation (planting nursery-grown trees or seeds and saplings gathered in the wild) and selective logging that took account of the growth rate of each species (Vidal y Soler 1874:25). To implement such a programme, foresters first needed to conduct surveys to determine which areas were to be logged and which were to be kept as forest (Vidal y Soler 1874:126). Foresters in the *Inspección general de Montes* may have been trained in German and French silvicultural principles but they adjusted European forestry to meet the particular conditions of tropical forests. The service began modestly in 1863 and grew to be a sizable agency (Nano 1953:13). The primary legislation the service operated under was the provisional forestry regulations of 8 February 1873 that Ramón Jordana y Morera, author of the most definitive account of forestry in the Spanish Philippines, considered expressed the most advanced scientific principles of the time. However, even he had to admit that they had been formulated with 'little knowledge on how best to achieve these aims and [in ignorance] of the customs and necessities of the inhabitants' (Jordana y Morera 1891:241–242). The forestry service continued to operate right up until the end of the Spanish colonial period though its effectiveness was hampered by a gradual reduction in the number of staff after 1881. Unfortunately, its historical legacy has suffered much from the fire of 1897 that destroyed its extensive library, forest maps, natural history collection, and the entire archives.[12]

There does not appear to have been any appreciable loss of forest cover due to campaigning during the American occupation of the archipelago between 1898 and 1907. Nevertheless, the new U.S. administration realized the extent of deforestation in the more populated areas of the archipelago (Ahern 1901:11).

12 LC, Pinchot Papers, box 586, 'Regino Garcia': 5–6.

Officials assessed this damage, however, more in terms of lost revenue and were struck by the amount of primary forest that still remained. 'The Philippine forests', wrote Worcester (1914:847), was 'like money in the bank'. By the beginning of the twentieth century, Americans were fast exhausting even their own prolific domestic timber reserves; foresters talked with urgency of an impending 'timber famine' and looked upon their new colonial possessions as a source of timber (Williams 2003:386–395). Worcester extolled the myriad uses for the different types of timber and non-timber products and a flow of publications detailing knowledge about the forest and the commercial usefulness of Philippine timber was a hallmark of the early decades of the American era (Miller 1911:97–111; Schneider 1916).

The period from 1898 to World War Two witnessed an enormous expansion of commercial logging around the archipelago with exports to markets in the USA, Japan, China and Europe (Schneider 1916:11). To meet this demand, operations were modernized with logging engines and railways largely replacing axes and carabaos (Collins 1901:82). Total annual lumber production rose from 94,000 to 2,500,000 cubic metres between 1901 and 1940. In particular, the volume of timber cut from lesser quality dipterocarp hardwoods (such as apitong, lauan and tangile) rose by over a 1,000%.[13] A moving timber frontier embraced the whole archipelago with the inclusion of extensive logging operations on Mindanao. A displaced and rapidly expanding population, moreover, sought new land to cultivate that was literally hacked out of the forest, with or without official consent. This encroachment only accelerated after the establishment of internal self-rule in 1935 (Severo et al. 1962:3). Again, the overall figures on forest cover do not properly reflect the magnitude of this loss. The inclusion of Mindanao into the colonial state in 1898, an island 'almost entirely covered in timber', effectively increased the extent of forest cover in the archipelago by a third. Yet by 1932 forest covered only 57% of the total land area, a fall of a further 9% since 1903 (Ahern 1901:11). The Great Depression of the early 1930s proved only a temporary respite.

The need for both timber and food intensified with the Japanese invasion and the harnessing of the archipelago's forests for wartime purposes. After a systematic bombing of Philippine defences, Japanese troops landed on 10 December, 1941 at several points on Luzon and rapidly overran the less well-equipped and much smaller American and Filipino forces. The Japanese were determined to make full use of the archipelago's forest resources not only the wood but also non-timber products such as rattan, resins, gums, tannin, oils and medicinal plants. However, timber production during the occupation was only a fraction of its pre-war level (Sulit 1947:35). Many factors explain this

13 Fischer (1925:64, 1937: 87–88, 238–239); Tamesis (1948: 320).

poor performance: wartime damage to mills, a shortage of draught animals, unrealistically low fixed timber prices, and the uncooperativeness of Filipinos, as well as widespread sabotage by guerrilla fighters (Sulit 1947:38–39). Despite plummeting production figures, however, the damage inflicted on the archipelago's forests during the occupation was substantial (Agoncillo 2001:607–635). Food shortages led to an unregulated explosion of slash-and-burn agriculture and to the clearance of much woodland (Severo et al. 1962:4). The Japanese military's demand for timber resulted in the extensive destruction of forest habitats including trees in national parks and forest reserves (Sulit 1947:40–41). Moreover, intense fighting occurred in many forested areas, first in early 1942 and again as a result of the scorched earth policy practiced by the Japanese Army in the months following the American landing in October 1944. The sheer scale of the destruction and the random indiscriminateness of the damage inflicted on the forest separate the Japanese occupation from all previous wartime experiences in the Philippines.

In the decades following World War Two, the wholesale commercial exploitation of the forest began in earnest, especially during the presidency of Ferdinand Marcos (1965–1986) (Broad and Cavanagh 1993:39–55). These years were characterized by cronyism, corruption, and corporate logging as the archipelago's forests were plundered on an unsustainable basis (Vitug 1993). Forest cover declined from 50% of land area to a little over 22% between 1950 and 1987 (Kummer 1992:56–57). The rapid decline in forest cover was not due solely to intensive logging (both legal and illegal) but also resulted in part from an expanding agricultural frontier to accommodate the archipelago's rising number of people. Population increased from 19.3 million to over 48 million between 1948 and 1980 (Kummer 1992:77). The decades following 1945 also saw the Philippines emerge as the major producer of tropical hardwoods in Southeast Asia for international markets. National log exports peaked at a high of 8.7 million cubic meters in 1969 before falling back to 5% of total exports by 1987 (Top 1998:64).

After the restoration of constitutional government in 1986, more progressive forest policies were introduced that included people-oriented social forestry programs. Provincial and local logging bans were instituted and forest reserves created. Reforestation projects and spontaneous tree-planting contributed to a rise of 0.7 million hectares in forest cover between 1988 and 2003. Since 1995, too, there has been greater official understanding that conservation and reforestation projects can only succeed with the direct involvement and participation of forest users. Community-based forestry management agreements that recognize local communities as forest managers are now acknowledged national strategy and, by 2006, over 6,000,000 hectares of forest lands were under some form of community forest management (Lasco and Pulhin 2006:48). Still, it may be rather premature to conclude that this trend represents a new phase in the relationship of humans to the forest. Indeed, such statistics often lack robustness and illegal logging and forest clearance continue apace in many areas, often cloaked in the rhetoric of community empowerment and poverty alleviation (Ploeg et al. 2011). By the end of the twentieth century, too, the Philippines had become a net importer of tropical hardwoods.

Conclusion: Deep Forestry and the Longue Durée

Adopting a 'deep forestry' approach may not really allow one to 'think more like a forest' but it does help reveal the true dynamics that govern the nature, composition and extent of the forest in the Philippines over the *longue durée*. In the first place, deep forestry offers a more inclusive understanding of the archipelago's woodlands: how they changed over time, who or what were the agents that precipitated such modifications, and what have been the consequences of their actions. Climate, soil, fire and animals were significant factors not only in determining the nature and composition of the forest but also in influencing human decision-making. As a result, any historiography of the forest needs to encompass an extended timespan that includes all these actors in its narrative. Seeing the forest as more than simply its constituent trees highlights its real value not only as a resource but as an important stabilizing element in the environment.

Taking a more biocentric approach shows how forest history is more than a mere chronicle of human actions. The sheer scale of forest clearance since 1946 has dwarfed most other considerations, but people have always acted with and within a forest ecosystem. The full impact of fire and white ant has been synergistic with human agency, while the soil and even the climate are nowadays also the product of a close interaction with people. In particular, climate change is set to become a major determinant of future agricultural yields as well as the health of the archipelago's remaining forests. Forest historiography needs to reflect these complex and dynamic processes. Finally, thinking more like a forest is a reminder to all that the forest has always acted as a community's first line of defence against the vicissitudes of climate and the extremes of seismicity. The Philippines are one of the most disaster-prone nations on earth with an average of 20 typhoons a year and an earthquake of varying magnitude occurring somewhere in the islands every day (Bankoff 2003). A deep forestry approach appreciates this contribution and the significant part the forest plays in mitigating the potentially destructive power of natural forces in an increasingly less predictable world.

CHAPTER 3

Breeding and Power in Southeast Asia

Horses, Mules and Donkeys in the Longue Durée

William G. Clarence-Smith

Introduction: Tibetan Origins

Horses leap out at the viewer from countless pictures and sculptures in Southeast Asia, and Peter Boomgaard has been a pioneer in developing their neglected history (Boomgaard 2004, 2007a). Horses clearly belong to the *longue durée* in Southeast Asian history, for they have been present for at least two millennia, linked structurally to war and elevated social status. In contrast, mules, the sterile offspring of horse and donkey, are hidden from view. They came later, spread less, and were employed for more humble pursuits. As for donkeys, they have been almost absent from the region (Clarence-Smith 2004b).

Horses are commonly assumed to have come from India and China.[1] Southeast Asian elites certainly absorbed much equestrian lore from these neighbouring civilizations.[2] India was also the probable source of prejudices against donkeys and mules (Kipling 1921:76–77, 203–205). For the animals themselves, however, archaeological analyses of equine bones are rare, and there are no DNA surveys. Linguistic evidence is tricky, while iconic, epigraphic and documentary records are sparse and fragmentary (Blench 2010; Wade 2009b).

Sifting through what evidence there is, Tibet appears to be the most likely origin of most Southeast Asian ponies, defined as horses measuring under 14 and a half hands (1.47 metres) at the withers (shoulders). These ponies entered the region via Yunnan, at a time when it was independent of China. However, they probably came ultimately from the equine corridor of eastern Tibet, with its cool dry climate and natural pastures. In contrast, grasslands in Southeast Asia and Yunnan were usually deliberately created and maintained by periodically firing vegetation.[3]

Horses may have emerged in Tibet in the second millennium BCE, at the same time as they entered China from the Eurasian steppe. They were certainly present by the first millennium BCE (Blench 2010:2; Bagley 2001:39). According

1 Boers (2007:52); Hoekstra (1948:39–40); Charney (2004:174).

2 Crawfurd (1971:154); Quaritch Wales (1952); Charney (2004:164).

3 Henley (2005a:8–9, 544–545); Reid (1988:118); Chapter 2 in this collection.

to one account, Tibetan ponies developed as half a dozen tough and sure-footed breeds, distinct from those of Mongolia, as natural or social selection encouraged adaptation to harsh conditions at high altitude.[4] As Tibetan ponies are smaller and less corpulent than those of Mongolia, an infusion of blood from across the Himalayas has been suggested (Kidd 1985:38, 47, 191).

The Chinese obtained Tibetan horses from early times (Bøckman 1991:177; Schafer 1985:62, 64–65, 76). The Qin, one of China's 'warring dynasties', expanded into the Tibetan borderlands of Sichuan in the fourth century BCE, partly to obtain such horses (Bagley 2001:39). Since then, the country of Kham, the south eastern Tibetan marches, has often been annexed to Sichuan, leading authors to refer to the 'wiry pony of Szechwan' (Schafer 1985:62). In reality, the great rice-growing bowl of Sichuan proper is not a land of equids (Gill 1883:88, 107; Davies 1909:222, 292–294).

Tibetan ponies probably diffused onto the limestone plateau of Yunnan, where archaeological evidence suggests that they were raised from the sixth century BCE (Wade 2009b:162). Dian bronzes from the independent state of the region, dating from around the second century BCE onwards, show horses ridden without stirrups. Han China obtained Dian horses as gifts and tribute, or through raids (Pirazzoli-t'Serstevens 1974:19–20). The later Nan Zhao kingdom continued to supply ponies to China from the late sixth century CE (Backus 1981:9, 30, 119). To be sure, the Chinese long preferred sturdier and larger Mongolian ponies to Tibetan 'river ponies' so called because they were shipped down the Yangzi (Coates 1994, 22–24, 59, 186). However, ponies from the south-western marches were a significant resource, especially when the steppe turned hostile to the ruling dynasty in China.[5]

The Diffusion of Tibetan Ponies across Mainland Southeast Asia

Ponies moved easily from Yunnan into the northern uplands of Southeast Asia (Luce 1960:326; Forbes and Henley 1997:83). Indeed, there were no ecological barriers, and a mosaic of Burmic, Tai, Austroasiatic, and Miao-Yao peoples lived on either side of the current border with China (Backus 1981). In the ninth century CE, horse breeding was especially developed in what are today western Yunnan and northern Burma (Wade 2009b:166–167; Luce 1960:326).

Archaeological evidence suggests that horses were being bred, and copied in figurines, on the Khorat Plateau of northeastern Thailand between 500 BCE

4 Peissel (2000:82–83, 127–130, 184, 235–241).

5 Backus (1981:163); Paul Smith (1991); Yang (2004:297–299).

and 400 CE (Wade 2009b:162). Horse breeding spread easily into Southeast Asia's Dry Zone, where a rain shadow effect lowers annual precipitation. Open savannahs provided rough grazing, extending over a discontinuous belt of plains and low plateaus from central Burma, to northeastern Thailand, Cambodia, and south-central Vietnam (Fukui 1999). Local fodder crops, such as sorghum in central Burma, were also well adapted for horses (Charney 2004:172).

Trade with India and China is unlikely to have had much effect on the bloodlines of Southeast Asia. The balance of equine trade always appears to have been in Southeast Asia's favour. For example, China obtained Yunnanese horses via northern Vietnam in the first millennium CE (Yang 2004:295–296). Even when horses moved the other way, they mainly consisted of a kind of Tibetan pony, as in the case of northeastern India.[6] A smaller version of the Tibetan pony also prevailed south of the Yangzi in China (Mundy 1919:III-1, 266, n.4; Moor 1968:189).

The hypothesis of a Tibetan origin for the great majority of horses in Mainland Southeast Asia is reinforced by morphological considerations. Although ponies naturally tend to become smaller and 'weedier' in hot lowland zones, they continue to display marked physical affinities across the whole region (Moor 1968:189). That said, DNA tests would be necessary to resolve scientifically the question of how populations relate to one another.

The Mule Frontier

The common small grey donkey, coming either from the steppe or from India, was ubiquitous in Tibet by the end of the nineteenth century, being raised by most farmers and herders (Bell 1998:44; Legendre 1913:194). Tibet also had larger donkeys, however, some of which were sent to the Chinese court as a gift as early as 654 CE. Mules were reported to be present in Tibet shortly thereafter (Schafer 1985:8, 76). In modern times, Tibetan riding mules were considerably larger than pack mules, reflecting the existence of large breeding donkeys, possibly imported from North China.[7]

Mules and donkeys followed horses onto the Yunnan Plateau from Tibet at an uncertain date, but the Mongol annexation of Yunnan, in the mid-thirteenth century CE, probably gave mule breeding a boost. The Mongols

6 Chakravarti (2009:156–157); Joshi and Brown (1987:308); Chandola (1987:76); Balfour (1871, II:613); Kipling (1921:181).

7 Teichman (1922:101, 221–222, Plate LXII).

garrisoned their newly conquered province with Inner Asian Muslims, who gradually evolved into Mandarin-speaking Hui communities, renowned for their mule breeding skills.[8] Mules were certainly common on the Yunnan Plateau by the late nineteenth century, and donkeys were present on a small scale (Pourias 1889:16). However, breeding hardly spread at all into Southeast Asia.[9]

Mules were more widely employed in Mainland Southeast Asia, as they plodded along the lengthy trails of the 'southern silk road'. These routes went back to the second century BCE, although human porters initially carried goods (Pirazzoli-t'Serstevens 1974:45, 63). Mules may have been employed in caravans by the sixth century (Giersch 2006:170–171, 174–179). They certainly travelled on Tibetan routes in the eighth century (Schafer 1985:8).

By the late nineteenth century, mules dominated the booming caravan traffic, with an average of five or six mules to one pony (A. Forbes 1987:28). Yunnanese muleteers, especially Muslim Hui, developed a spider's web of routes, radiating out in all directions from three main entrepôts in Yunnan. Most termini were river ports, although caravans followed the un-navigable Salween to the sea at Moulmein. The muleteers also shifted bulky goods to yaks on the edges of the high Tibetan Plateau (A. Hill 1998; Forbes and Henley 1997:79–89).

Southeast Asian dependence on imports of mules from Yunnan and Tibet persisted (Clarence-Smith 2004b; Younghusband 1995:77–78). Burma appeared as an exporter to British India and Malaya in colonial times, but these were re-exports of Tibetan and Yunnanese equids (Courtellemont 1904:173–179, 242). The reluctance of most Southeast Asians to engage in mule breeding themselves remains to be explained.

The Spread of Horses in Maritime Southeast Asia

The same broad type of pony can be found all across Southeast Asia, as far as Timor.[10] As dates of introduction are later than for the mainland, this suggests that the equine stock of the islands came mostly from the mainland, and was moved in short 'island-hopping' trips on small boats.

The famous sculptures in the temple of Borobodur in south-central Java, which date from the eighth and ninth centuries CE, seem to portray real local

8 Forbes and Henley (1997:29–31, 38, 53, 79); Wang (1996:76, 117–125, 156).

9 Scott and Hardiman (1900–01, II:1, 200–202, 205, II-2, 68); Younghusband (1995:77–78); Nguyen (1965:141).

10 Moor (1968:189–190); Goodall (1973:167–168, 177–181); Kidd (1985:193).

ponies, rather than idealized Indic types (Wade 2009b:167). Some horses may already have been in Java in the seventh century, but the evidence is equivocal (Knapen 2001:294; Burkill 1966, I:1216). By the thirteenth century, 'hill ponies' were employed for riding inland in Java (Chao Ju Kua 1911:77).

A veritable 'cowboy country' developed in Nusa Tenggara (the Lesser Sunda Islands), an archipelago that becomes increasingly dry as it approaches Australia, and has the highest concentrations of horses in Southeast Asia. Epigraphic evidence shows that they were bred in Bali by the tenth century (Lombard 1967:89; Atmodjo 1985:31–32, 48). Evidence from bones has them present in the island of Timor by about the fifteenth century.[11]

In other parts of Maritime Southeast Asia, horses only spread after Europeans had arrived, but they were of the local type. Archaeologists have excavated no horse bones before the Spanish conquest of the Visayas in the late sixteenth century (Nishimura 1999:35). Nor is there any evidence for their presence, there or in Luzon, in other sources.[12] Moreover, words for horse in the languages of these parts of the archipelago derive from Spanish or Portuguese (Crawfurd 1971:154; W. Forbes 1976:16). Sulawesi was a similar case.[13] Nevertheless, Europeans imported ponies from neighbouring islands, or from the mainland, and not from further afield (Wernstedt and Spencer 1967:211).

Abundant exports of ponies from the eastern archipelago contrast with the failure of Mainland Southeast Asia to become self-sufficient. In the early sixteenth century already, Bali, Lombok and Sumbawa were supplying Java with numerous horses (Pires 1944, I:202–203). By the nineteenth century, Nusa Tenggara and South Sulawesi exported many ponies, not only to Java, but even to Mauritius in the western Indian Ocean.[14] The Philippines developed separate patterns of horse exports, to south-eastern India in the eighteenth century, and to south-eastern China in the nineteenth century (Quiason 1966:73, 105; Coates 1994:15–16, 29).

The northern half of Sumatra was another exporter, albeit on a smaller scale. There may have been horses here already in the third century CE, although the identification of place names is problematic (Chakravarti 1999:205). By the twelfth century, Chinese texts suggest that exports had begun (Chao Ju Kua 1911:72; Wade 2009b:170). The mountainous northern half of the island certainly developed a vigorous equine complex over time (Boomgaard 2004:213–216).

11 Ramos and Rodrigues (1980); Ian Glover, personal communication, 2004.

12 Pires (1944, I:133–134); Balfour (1971, II:614); Junker (1999:327).

13 Pelras (1996:78–79, 99–100); Henley (2005a:69); Crawfurd (1971:277).

14 Parimartha (1995:137–138); Clarence-Smith (2002, 2004a:57–64).

In the seventeenth century, Aceh sent 'small, yet sprightly' ponies to south-eastern India (Lombard 1967:89, 110, 221; Ito 1984:378). In the nineteenth century, 'Batak' and 'Achin' ponies crossed the straits to the towns of the Malayan Peninsula (Balfour 1871:613; Kreemer 1922–23, I:512–513).

There were limits to the penetration of horses into Maritime Southeast Asia, even after the Europeans had arrived, with ponies absent from certain apparently suitable uplands. Notable examples were the interior of the Malay Peninsula, the southern half of Sumatra's Barisan range, and the stupendous mountains of New Guinea. Most of the vast Borneo plateau was similarly void of horses, albeit with minor exceptions (Moor 1968:190).

Long-distance Maritime Imports and Exports

A few horses were brought by sea over long distances, from the first millennium CE onwards, notably as gifts from Chinese emperors to loyal tributaries. Southeast Asian envoys to China in the seventh century specifically requested horses. Although it is not clear whether such requests were always heeded, Champa received fairly regular presents of horses from Song emperors from the late tenth century (Wade 2009b:165–168; Burkill 1966, I:1216).

However, exports of horses from China to Southeast Asia were tightly controlled. In 1175, officials of the central court upbraided a Fujianese trader for promising to sell horses to Champa, as 'China's horses have never been permitted to be sent beyond the borders' (Wade 2009b:169). Chronically unable to obtain sufficient horses to meet the empire's needs, China did not wish to send large numbers to 'southern barbarians'. Moreover, horses could not easily be smuggled unnoticed onto a junk.

The flow of horses went mainly in the opposite direction, as tribute to China. Vietnam and Yunnan were amongst such tributaries from early in the first millennium CE. After the Ming dynasty had seized power in 1368, it faced the hostility of the steppe. Horses then came to China more regularly from all over Southeast Asia, in small numbers by sea, and in larger numbers overland (Ptak 1991; Wade 2009b:168, 173–174).

India was also chronically short of horses, and thus sent no more than a trickle to Southeast Asia. As early as the third century CE, horses from India may have reached the Straits of Melaka and Cambodia, although the destination could equally have been Tamilnadu.[15] The traffic to Southeast Asia was

15 Chakravarti (1999:205, 2009:151–152); Wade (2009b:164–165).

occasional, with no large numbers ever dispatched, and it was more than counter-balanced by Southeast Asian ponies entering southern India.[16]

With the arrival of Europeans in the sixteenth century, potential new sources of imports opened up, but with few practical results. Portuguese officials, wishing to befriend Southeast Asian rulers, occasionally gave them fine horses, some of which were described as European (Lombard 1967:228–229). However, the Portuguese were more likely to donate Arabs and Persians, of which they shipped large numbers to India (Loureiro 2009). The Dutch VOC also initially conferred Arabs and Persians on its protégés, before cutting costs by breeding horses on islands off the north coast of Sri Lanka (Boomgaard 2004:212–213).

The Spaniards hardly had a greater impact. Very few horses came from New Spain, as it was so difficult to keep them alive crossing the vast Pacific Ocean (Diaz-Trechuelo 1963:203). Ponies from China and Japan were more significant, as the Spaniards found ways of eluding prohibitions on exports. In any event, the standard Southeast Asian pony soon prevailed, possibly obtained from the Muslim foe in Mindanao and Sulu (Bankoff 2007a:85, 191).

The embryonic nature of mule breeding in the Philippines was perplexing, as the Spaniards transformed the economy of the Americas with these animals. They brought a few donkeys and mules from New Spain, but with no significant impact (Diaz-Trechuelo 1963:203). China was more important as a source of donkeys and mules, but again with no apparent effect on local breeding (Blair and Robertson 1903–07:xvi, 90, 180).

A similar lack of donkeys and mules characterized the rest of Maritime Southeast Asia, even though a word for donkey was generally known (Crawfurd 1971:25). Donkeys appeared occasionally in Indonesia, but as oddities. They were reportedly in a royal procession of 1359 in Java, in the company of camels, perhaps imported to demonstrate royal power (Robson 1995:68, 128). In Batavia in 1709, 'a train of slave women on donkeys' followed Abraham van Riebeeck, the Dutch governor-general, in a similar manifestation of authority (J. Taylor 1983:57).

New Breeds in Southeast Asia to c. 1800

The genetic consequences of even quite small amounts of new blood could have been major. Deliveries of horses from China probably consisted of Tibetan

16 Wade (2009b:172); Chao Ju Kua (1911:72); Ito (1984:378); Quiason (1966:73, 105).

and Mongolian ponies, although 'divine' Central Asian horses of Fergana may have reached favourite allies. Ships would also have brought tough Indian crossbreeds from Kutch and Sind, fine Arabians, and large Türkmen (Turcomans) from Persia. Western stock, whether from Europe or the Americas, was rare.

In reality, it is difficult to detect much impact from this outside blood on the ponies of Mainland Southeast Asia, because larger and 'finer' horses acclimatized badly in tropical conditions. Early modern Thai kings made sustained efforts to obtain Middle Eastern horses, and, failing that, horses from Java, but there is no indication that this led to self-reproducing new breeds (Pombejra 2007). Less is known about Burma, Cambodia or Vietnam, but again, no new type of horse can be seen to have emerged.

In Maritime Southeast Asia, new breeds did appear, of which the largest was the Priangan (Preanger) horse of western Java. It emerged from the late seventeenth century, as a result of crossbreeding with Middle Eastern horses, a process that was actively encouraged by the Dutch from 1736 (Boomgaard 2004:218–219, 223). As Arabs are small, only just qualifying as horses rather than ponies, the Persians were probably of the big bony Türkmen type. By the 1890s, Priangan horses might reach 15 hands, but they had weak bones and were 'razor-chested' (Worsfold 1893:220). Although this was the tallest local horse in Southeast Asia, one observer thought that it 'otherwise cannot be highly praised' (Cabaton 1911:119). Moreover, the Priangan horses tended not to breed true (Barwegen 2005:46–47; Sibinga Muldcr 1927:321).

The Kedu horse of south-central Java was quite different, as it was the result of careful selection within local equine stocks by Javanese aristocrats from the early eighteenth century (Boomgaard 2004:220). Though smaller than the Priangan horse, it was much better proportioned (Cabaton 1911:120; Barwegen 2005:47). One specialist of the 1930s was bowled over by the beauty of this 'miniature horse' (Kay 1939:24–25).

The most famous Maritime Southeast Asian breed was the Sandalwood, which probably emerged in early modern times. Variously said to originate in Sumbawa or Sumba, it was known as 'the handsomest and best breed' and 'the Arab of the archipelago'.[17] In accordance with this sobriquet, it has been alleged that there must have been Arab blood in its veins (Kidd 1985:193). There is no hard evidence to substantiate this view, however, and selective breeding may have been more important for its development.

17 Boers (1997, 2007); Barwegen (2005:49–50); Crawfurd (1971:53, 153–154, 375, 420).

Horses and Southeast Asian State Formation

A connection between horses and political power emerged, in part because of the speedy communications that these animals afforded. Mounted units also made armies more mobile, and cavalrymen were a particularly formidable force on the battlefield before the rise of firearms. Even after the introduction of firearms, cavalry continued to play an important role in scouting, foraging, flanking, probing, and the pursuit of fugitives (Boomgaard 2004, 2007a; Charney 2004:Ch. 7).

It is thus unsurprising that Southeast Asia's earliest large states often had capital cities located close to supplies of horses, as in Burma. The capital of Pagan, which flourished from the ninth to the thirteenth century CE, was in a dry area, with good access to horses. The Chinese customs official Zhao Rugua (Chao Ju-kua), who compiled information about the twelfth and early thirteenth centuries, reported on 'a great plenty of horses', ridden without a saddle (Chao Ju Kua 1911:58). Of 185 mentions of animals in Pagan inscriptions known to scholars around 1940, 15 referred to horses (Luce 1960:370, 373–375). Tomé Pires (1944, I:102, 111), in the early sixteenth century stressed the importance of both horses and elephants to the power of the successor state of Ava, whereas he noted only elephants for the rival state of Pegu to the south.

Place names further indicate that specialized horse breeding areas extended from the city of Pagan eastwards to Kyaukse (Frasch 1996:180). Indeed, the uplands of the northern Pegu Yoma range, especially the slopes of mystical Mount Popa, long remained famous for raising legendary horses (Scott and Hardiman 1900–1901, II:2, 263).

As many historians have noted, the capital of the great empire of Angkor, which flourished from the ninth to the fourteenth century, was not located directly in the 'rice bowl' of Battambang. Carvings show many horses, ridden without stirrups, as well as elephants.[18] Inscriptions also refer to horses (Quaritch Wales 1952:88). That said, Zhou Daguan (Chou Ta-kuan), a Chinese envoy in Angkor in 1296, stated that these horses were 'very small' (Chou 1967:35).

The confederation of Champa, in what is today central Vietnam, had few rice fields, but disposed of numerous horses, as well as elephants (Wade 2009b:165). Champa may have bred the smallest horses in all of Southeast Asia, for Ma Huan declared them to be as small as donkeys (Ma Huan 1970:81). Nevertheless, they were plentiful, for Tomé Pires (1944, I:113) noted of the Cham

18 Frédéric (1981:322–324); Jacq-Hergoualc'h (1979:75–77, 84, 143–144, 149–161).

in the early sixteenth century that 'they all have horses'. Indeed, by the late eighteenth century, horses were running almost wild in the mountains of what is today south-central Vietnam (Li and Reid 1993:125).

The Vietnamese of the Red River, blessed with rice, horses and elephants, developed a precocious horse culture, based in part on easy access to Yunnanese horses (Yang 2004:295–296). China ruled this area from 111 BCE to 939 CE, integrating it into a civilization where equids were part of daily life. Vietnamese tombs of the Han era contain model farms, made of ceramics, with horses among the domestic stock (K. Taylor 1983:50; K. Hall 1992:263). In the mid-ninth century, mountain chiefs from the fringes of the Yunnan Plateau bartered horses and cattle for lowland salt. When a senior Chinese official took control of salt producing villages and altered the terms of trade, the inhabitants called on Nan Zhao for redress.[19] In early modern times, northern Vietnam was described as 'a land of many horses', with breeding an upland speciality (Pires 1944, I:115; Nguyen 1965:141–142).

The relation between horses and centralized states may have been greatest in Java, where elephants were not present in the wild, though a few were imported by sea. While states of the 'Malay World' often rested on the control of sea-lanes, this was less true of densely populated Java, where cavalry was crucial to political centralization (Quaritch Wales 1952:59–61; Pires 1944, I:193, 227). The capital of Majapahit was situated close to the numerous horses that Tomé Pires (1944, I:198) reported as grazing in the eastern part of the island.

The adoption of firearms from the fifteenth century had a somewhat contradictory impact on horse-borne warfare. Capital cities were often moved to seaports, where the new weapons could most easily be imported. However, firearms also made horses more useful than elephants, as the latter presented easier targets and tended to panic in the din of battle (Charney 2004:160–163, 183–185).

The spread of new Abrahamic faiths may also have contributed to raising the profile of horses. For Muslims, they were unquestionably noble beasts, whereas elephants had a somewhat troubled reputation, having been employed to attack the Prophet. Elephants were certainly part of Muslim royal courts and armies in South and Southeast Asia, but they tended to make way for horses over time (Clarence-Smith 2004c). Christians found elephants strange and hard to handle, rather than religiously tainted, while sharing the Muslims' enthusiasm for horses.

---

19 K. Taylor (1983:239–240); K. Hall (1992:266); Wade (2009b:169).

Modern Western Colonialism and Horse Breeding

The modern Western colonial occupation of Asia and Africa depended greatly on equids in military terms. 'Flying columns' consisted of dismantled mountain and machine guns on the backs of ponies or mules, squadrons of light cavalry, and a mounted baggage train able to keep up with the first two forces (Callwell 1990).

Colonial officers and officials, imbued with notions of Western scientific superiority, thus sought to transform the breeding of Southeast Asian horses, but they met with almost complete failure. Imported horses refused to eat local fodder, reproduced poorly, and died quickly, and their crossbred progeny suffered from similar defects. Moreover, they were subject to the dwarfing that occurs naturally among horses in tropical conditions (Bankoff 2007b:105–106, 109–111). One particular problem in Southeast Asia was that bamboo was a significant fodder, which imported stock would not eat (Younghusband 1995:76, 144; Rémond and Winter 1944:29).

The Spaniards imported Arab stallions into the Philippines in 1858, to produce artillery horses, but they actually died of starvation, due to a lack of suitable fodder (Bankoff 2007b:108–111). Nor did the Americans do much better with imports of American and Australian breeding stock, because of a combination of disease and feeding problems (Elliott 1968:346; Foreman 1906: 336–337).

Although the French claimed more success in Indochina, this is open to doubt. Over two generations, Syrian Arabs and Tunisian Barbs were crossed to produce horses that were three quarters local by blood, slightly larger, and allegedly retaining the endurance and resistance of local animals (Teston and Percheron 1931:916–917, 919). However, the British judgement was that French crossbreeding since the end of the nineteenth century had failed, including attempts with horses imported from Java and Burma (*Indo-China* 1943:288–289). Nor were efforts with Australian imports in neighbouring Thailand crowned with any greater success (Thompson 1941:332).

Cultural misunderstandings also occurred, for example in the Chin area of Burma. Local people refused to have their mares serviced by a prize stallion provided free by the Frontier Force. They explained to a perplexed British official that they were inexpert riders, and that they were 'not keen to produce a stronger and more fiercely intractable breed' (Stevenson 1970:60).

Colonial efforts in the Malay world fared no better. The British in Penang mated Arab or Persian stallions with Sumatran mares around 1830, but soon gave up these experiments (Burkill 1966:1217). The Dutch were scarred by the Java War, as the metropolitan government had decided to abolish cavalry and horsed artillery in 1825, just as the island was about to explode into a destructive rebellion that lasted till 1830. An initial shortage of horses nearly resulted in disaster for Dutch arms (Heshusius 1978:8–11). A pre-existing stud in Priangan was thus reactivated (Payen 1988:100). Persians, Arabs and Sandalwoods were then crossbred with local mares throughout the nineteenth century, with Australian Walers added in the 1920s. However, results were disappointing (Barwegen 2005:44–45). Indeed, attempts by officers and racing enthusiasts to breed large horses were criticized as a waste of money, which would have been better spent importing ponies from Nusa Tenggara. Small horses were normal in the tropics, it was argued, and it was pointless to struggle against nature (Sibinga Mulder 1927:321–322).

Persistent efforts to 'improve' the horses of Nusa Tenggara itself, with imported Arab and Australian stock, accomplished little. That said, private initiatives on the part of Hadhrami Arab horse traders and local rajas seem to have been somewhat more effective than those undertaken by government studs.[20]

Incapable of changing Southeast Asian horses much, racing enthusiasts and governments fell back on importing Australian Walers. These fine horses, bred in a semi-tropical climate, were sold from South Africa to Japan by the end of the nineteenth century. By 1941, almost all the horses required for military, police, and parade duties in Southeast Asia were Walers, as well as those dominating race-tracks and polo grounds (Yarwood 1989). Local ponies were left with the drudgery of transport work, and occasional participation in agricultural tasks.

Modern Western Colonialism and Mule Breeding

Colonial powers were no more successful in enhancing mule breeding. The Dutch were the most refractory to the possibility, having no significant experience of these hybrids at home, and only minor exposure to them in their Caribbean colonies. Despite the appearance of donkeys in Java in 1709, as noted above, there are no indications that the Dutch ever thought of fostering the breeding of mules in Indonesia.

Iberians had the necessary skills and inclinations, and yet proved unable to act. A proposal to import Spanish mules for the armed forces of the Philippines,

20 Clarence-Smith (2002); Parimartha (1995:206); Patji (1991:165); Ormeling (1956:115); Hoekstra (1948:45–46); Sibinga Mulder (1927:323–324); Wielenga (1926:16, 35–36).

in 1883, reflected a failure to breed them locally (Bankoff 2007b:109). In that decade, John Foreman encountered a grand total of five donkeys in his peregrinations around the Philippines (Foreman 1890:388). The Portuguese, who shared a common mule culture with the Spaniards, never even seem to have tried to raise mules in East Timor.

The American victors of 1898 in the Philippines were probably the world's greatest mule breeders at the time, but they did no better than their Spanish predecessors. The Americans brought numerous military mules to the Philippines, at great expense, and therefore imported 10 jack (male) donkeys in 1904 to breed mules locally in the government's experimental stock farms (Montemayor 1952:279). Private companies sought to do likewise.[21] As none of these initiatives came to fruition, infantry officers concluded in 1927 that it would be better to recruit 'native ponies' as pack animals, rather than to import the 'time-honored Army mule' from America.[22]

The French, heirs to a similar glorious tradition of mule breeding, achieved even less in Indochina. They also imported many mules, to impose their rule in the mountains of northern Indochina, obtaining them from China, Algeria and France (Teston and Percheron 1931:914). Despite their military reliance on these animals, notably to transport mountain guns, they did not develop the embryonic Vietnamese breeding sector in their colony.

The British had little experience of mules at home, but they had built up some credentials in the Punjab in northwestern India, and had high hopes for Burma, after its final conquest in the 1880s. The Army of India's mountain batteries suffered from a chronic deficit of mules, which had to be filled by expensive imports from Argentina, the United States, and elsewhere (Tegetmeier and Sutherland 1895:135–137). Persian jack (male) donkeys were therefore provided free for stud purposes in the southern Shan States of Burma. However, for reasons that are not stated, 'mule-breeding did not prove popular'.[23]

Malaya was not a target of British mule breeding aspirations, although mules were present. They were reportedly employed before 1917 to transport tin ore from an inland mine in Pahang, Malaya (Ledesma et al. 1997:134). An Indian mule battery, stationed in Singapore, mutinied in 1915 (Dijk 2009:124). Such mules were almost certainly imported from China and India respectively.

21 NARA, RG 350 (Bureau of Insular Affairs), 1914–45, Box 272, 1214–54, No. 3081, 'I.X.L. Cattle Co. Inc.', 7 December 1923.

22 NARA, RG 350 (Bureau of Insular Affairs), 1914–45, Box 422, 2392–149, War Department, Note, 1 June 1927.

23 *Imperial Gazetteer* 1907–1909:xxii, 258.

A shortage of mares would have inhibited local mule breeding, even if Pashtun (Pathan) immigrants might have introduced donkeys into Kelantan.[24]

Conclusion

Horses constitute a living testimony to relations between Southeast Asia and Tibet in the *longue durée*, suggesting that scholars may have underestimated this dimension of the region's history. While Buddhist contacts have received a little attention, material exchanges with Tibet, especially significant for Mainland Southeast Asia, remain largely unexplored. Yunnan has acted as a kind of screen, portrayed only as a conduit to China, whereas it also led to the grasslands of the 'roof of the world'. Further research on animals in Southeast Asia, following in Peter Boomgaard's footsteps, might develop the wider implications of being located between Tibet and Australia, two of the greatest pastoral regions in the world.

Conversely, historians have tended to overestimate the impact of horses brought by sea from outside Southeast Asia. These beasts were few in number, and left little genetic imprint. Where successful new breeds appeared, they were mainly the result of selecting and managing existing animals, and not of crossbreeding with imported stock. Indeed, by the second half of the twentieth century, experts had come round to the view that importing foreign horses into the tropics, or artificial insemination from such sources, was an illusory method of ameliorating breeds. It was much more effective, if slower, to select within 'native types', and to improve feeding and management (Tempany and Grist 1958:224–226; Barlow 1990:46–47).

The arrested development of mule breeding in Southeast Asia is a final topic that requires more attention, though it is always difficult to research a negative problem. Southeast Asia is not alone in this situation, for the same question could be asked of Japan and Korea, or indeed of Britain and northern Europe. Cultural and religious prejudices against donkeys, and against hybridizing animals of different species, may well hold the key to this puzzle (Dent 1972).

24 Khoo Kay Kim, personal communication, 24 October 2001.

CHAPTER 4

Under the Volcano

Stabilizing the Early Javanese State in an Unstable Environment

Jan Wisseman Christie

Introduction

Over the millennia the impact of human beings on the landscape and environment of Java has been profound. But the converse has also been true: the environment of the island – both climatic and geological – has shaped the lives of those who dwell there. Given the proximity of Java's active volcanoes to densely populated areas of the island, the devastation they have caused has at times been significant. Yet, possibly because of the relative geological quiescence of the twentieth century, many historians have tended to discount the role played by eruptions and earthquakes in shaping the early political and social history of the island.[1]

Memories of the destructive force of the volcanoes are, however, preserved in Javanese chronicles, and these were often invested with political significance. The massive eruption of the volcano Merapi at the heart of central Java late in 1822, perceived as an omen of the coming of the mythic 'Just King' in a time of famine and oppression, helped to precipitate the Java War of 1825 (Carey 1986:131). Eighteenth-century central Java perceived the ash-rains and eruptions of the volcanoes Merapi and Prahu in the 1760s and 1770s as signs heralding collapse of the state (Ricklefs 1974:186). The *Babad Tanah Djawi* states that the death of the great seventeenth-century king Sultan Agung was marked by Merapi's rumblings (Ricklefs 1974:18), and, according to the *Babad ing Sangkala*, that volcano's major eruption of 1672 not only killed many outright, but presaged far greater political and military calamities (Ricklefs 1978:181). The fifteenth century *Pararaton* lists eight eruptions and one major earthquake between 1311 and 1486, most invested with significance in the world of men (Phalgunadi 1996). These histories suggest that for centuries the Javanese had suffered periodic bouts of geological instability, and that throughout that time, supernatural agency was ascribed to them.

Investigating the geological upheavals of earlier periods is more difficult. No chronicles have survived from the first millennium, and the courtly *kakawin*

1 See the article by Anthony Reid in this volume.

 | DOI 10.1163/9789004288058_005

literature of the time, based on Indian originals, was too distanced from events in Java to provide more than a hint that eruptions and earthquakes were perceived to be the results of divine activity. Archaeology and geology provide some clues to the geological stresses of the times, but few concerning their impact on the people living in close proximity to the volcanoes. Of the records produced in Java during later first millennium, the only survivors are inscriptions written on stone and metal, most of which were produced by the state of Mataram. Since almost all of these inscriptions are either religious documents or tax charters, the information they provide is indirect. Much, however, can be gleaned from the changing patterns of political and religious activity they record, and these changes suggest that the state of Mataram, and its people, were forced to cope with even greater geological disasters than were their successors. The fact that the state survived says much for its resilience.

Early Mataram and Its Century of Turbulence

Mataram was one of the great historic polities of maritime Southeast Asia. Its wealth rested upon a combination of agricultural abundance and extensive trade (Wisseman Christie 1998, 1999). The number and size of its stone-built temple complexes, and the quantities of gold, silver and textiles that were transferred at ceremonies marking the benefice grants that were made for the support of these religious foundations attest to this wealth.[2] The state's agrarian abundance came at a price, however, since it owed the richness of its soils to the active volcanoes that surrounded its most densely populated areas. Although Mataram survived for more than three centuries, it did not remain the same. Over that period it first expanded rapidly; then met with disaster and contracted sharply, moving its centre several hundred kilometers to the east; and then re-stabilized for a time before finally breaking down.

The state of Mataram was founded early in the eighth century, probably in the northern part of the Kedu plain of central Java between the volcanic Dieng massif and the volcanoes Merbabu and Merapi. A few inscriptions of the later ninth and early tenth centuries identify the founder of the state as *ratu* Sañjaya, whose reign began in 716 CE (Damais 1951:42–63). Given the paucity of data, little can be said at this point about the earliest phase of the state. However,

2 Gifts to the hundreds of witnesses attending a series of benefice ceremonies held in the 870s and 880s for the support of one temple, for instance, received almost 1.75 kilograms of gold, 0.25 kilograms of silver, more than 800 pieces of cloth, large numbers of iron tools and other items, and substantial amounts of food. See Wisseman Christie (1996: 275–276).

most of the reigns of the first century and a half following Mataram's foundation seem to have been relatively long and successful. By the end of the eighth century the state had expanded to the south and east in central Java, encircling Merapi and Merbabu, to encompass the area around modern Yogyakarta and Prambanan (Wisseman Christie 2001:32–38). The first break in this series of stable reigns seems to have occurred in 847 CE, and it initiated a century of intermittent political upheaval. Piecing together information scattered amongst the surviving inscriptions – particularly that provided by the Wanua Tĕngah III inscription[3] – it is possible to assemble a preliminary outline of the political events that occurred between 847 and 948 CE.

This period of repeated instability was initiated when *rake* Pikatan *dyah* Saladū mounted the throne in 847 CE. His tenure on the throne was contested, and his reign was relatively short. When he died in 855 CE, the throne passed to *rake* Kayuwangi *dyah* Lokapāla. *Rake* Kayuwangi apparently oversaw a rapid expansion of the state to the east, taking control of the heartland of east Java by about 860 CE. *Rake* Kayuwangi's death early in 885 CE, however, precipitated a period of increasing turbulence. Three rulers succeeded to the throne within less than two years, before the state descended into civil war (Wisseman Christie 2001:38–50). Finally, late in 894 CE, the state was reunited when *rake* Wungkal Humalang took the throne. Following his death in 898 CE, *rake* Watukura *dyah* Balitung mounted the throne. In about 910 CE Balitung died and was succeeded in a regular manner by his heir-apparent, *rake* Hino *pu* Dakṣa. Dakṣa was, in turn, succeeded in about 919 CE – again apparently peacefully. However, the next ruler, who came to the throne in around 927 CE, was displaced before the end of that year. The new ruler held the throne for only about a year before being succeeded by his heir-apparent, *rake* Hino *pu* Siṇḍok, who remained on the throne until about 948 CE. The significant aspect of this last transition, aside from the brevity of his predecessor's reign, is the fact that from the time Siṇḍok took over, in 929 CE, Mataram's inscriptions were issued solely in east Java. Although the state continued to exist for more than a century, it issued no inscriptions in central Java after 928 CE. Something drastic appears to have happened to cause the state to abandon its established core region. The fact that no other political power seems to have filled the void in central Java suggests that the cataclysm was natural rather than man-made.

The surviving inscriptions, and their changing contents and patterns of distribution, provide an insight – though admittedly an oblique one – into the stresses, both man-made and geological, which helped to shape the political

3 This inscription provides a dated king list for part of this period (Kusen 1988–9; Wisseman Christie 2001).

and religious ideology of this early Javanese state, and which provided, to a degree, a template for its successors. The nature of these stresses, and the state's responses to them, form the focus of the discussion below.

Evidence of Geological Stress

As noted above, the challenges faced by Mataram between the mid-ninth and mid-tenth centuries were great enough to cause considerable – though episodic – political instability. This turbulence reached a peak in 928 CE, when the state's political and economic heartland in the densely populated and agriculturally rich lowlands surrounding Merapi in central Java was apparently abandoned, along with the majority of its monumental temples and religious foundations.

A major eruption, or series of eruptions, of Merapi, at the heart of the central Javanese portion of the state is the most probable cause of this turmoil, and physical evidence seems to confirm this suggestion.[4] Deep layers of *lahar*, or volcanic mud, built up in relatively rapid succession in low-lying areas to the south and southeast of the mountain.[5] Temples such as Sambi Sari and Kedulan, built during the ninth century in these low-lying areas, have been discovered buried under almost seven metres of *lahar*. These volcanic mud flows may also account for the large number of unrecovered hoards of small valuable objects – gold and silver currency, jewelry, military accoutrements, salvers and bowls, statues, ritual items and so forth – that have been found in the same region. The most spectacular of these finds, the Wonoboyo treasure, appears to comprise the entire contents of the strong room of a branch of the royal family.[6] Areas of central Java to the west of the mountain seem to have

4 There is some debate concerning the exact timing and nature of this volcanic activity, arising from an early eleventh century date suggested by R.W. van Bemmelen and some of his predecessors (Bemmelen 1956), based upon what was then known of the history of early Java. Given difficulties in precision of geological dating, more recent studies have tended not to question that original dating (Newhall et al. 2000; Gomez et al. 2010).

5 There appear to have been at least five separate flows of volcanic mud, possibly similar in the pattern to the more recent build-up of volcanic deposits on parts of the Caribbean island of Montserrat: the surface of each layer of *lahar* in central Java appears to have been exposed for long enough to become weathered, but most not for long enough for significant plant colonization and soil formation to have occurred (Newhall et al. 2000: fig. 4/E1; personal observation during field surveys, 1975–1999).

6 For the Wonoboyo treasure, see Kusumohartono (1993) and Wahyono (1994); for other currency finds, see Wicks (1986) and Wisseman Christie (1996).

been affected at much the same time, apparently by air-borne ejecta: the ninth- and early tenth-century occupation layer at Borobudur, to the west of Merapi, is sealed by up to half a metre of volcanic ash.

This major eruption was not Merapi's first during this period. The sealed occupation layer at Borobudur is interrupted by three or four narrow bands of ash, indicating that there had been at least three separate, smaller episodes of volcanic activity in the later decades of occupation at the site (Wisseman Christie 2001:47). Damage caused by accompanying earthquakes may explain not only the half-completed reliefs on the base of Borobudur – interrupted when the sagging monument was rapidly shored up by encasing the base with a broad, stabilizing foot – but also the fact that most of the other temples of central Java were apparently under continual repair and reconstruction until the point when most were seemingly abandoned. The repeated shoring-up of temples suggests that problems caused by periodic earth tremors may have been significant.

Considering the degree to which political instability in central Java during later centuries is known to have been provoked or exacerbated by eruptions of Merapi and accompanying earthquakes, it would not be surprising if a far more catastrophic series of volcanic eruptions and damaging earthquakes caused considerable turmoil during the later ninth and early tenth centuries. A more detailed review of the inscriptions produced in central and east Java during this period provides some support for this supposition. The inscriptions also provide clues to the responses of the early Javanese to the geological disturbances of the times.

Religious Responses to Geological Stress

Several important strands can be identified amongst the ideological responses of the early Javanese to periods of instability. The rulers who mounted the throne of Mataram immediately following, or within a few years of, periods of turmoil – and thus presumably facing problems of consolidating their hold on the throne – tended to issue large numbers of tax charters endowing religious foundations. The most notable of these rulers were *rake* Kayuwangi, Balitung and Siṇḍok during the ninth and tenth centuries, and Airlangga early in the eleventh century. The nature of the religious foundations they endowed, however, changed over this time. The surviving inscriptions and religious structures suggest that, by the middle of the ninth century, religious activity supported by the state had already begun to shift from Buddhism – dominant in the lowlands during the later eighth and early ninth centuries – to forms of

Śaivism focused largely upon the slopes of volcanoes close to the heart of the state in central Java, and the volcanic uplands of Dieng and Ungaran in central Java and of Malang in east Java. Although existing Buddhist foundations were periodically re-endowed, most of the new foundations and endowments of the later ninth century were at least nominally Śaivite in nature. The Indian god Śiva found a ready niche in the fertile, volcanic interior of Java, as the ascetic god of destruction and re-creation, lord of fertility and of ghosts, associated with fire, mountains and the source of sacred water (Daniélou 1964:16,188–231).

By the beginning of the tenth century boundaries were becoming increasingly blurred between Śaivite cults (including those focused on Agastya and Gaṇeśa) and the cult of deified ancestors – apparently linked to the spirits of the volcanoes – and to water temples.[7] Religious activity also became increasingly focused upon ascetic and tantric practices (Zoetmulder 1968:287–288), and the major gods – both Hindu and Buddhist – were increasingly merged with each other. The general trend appears to have been away from the religious eclecticism of earlier centuries, and towards religious syncretism.

Particularly prominent amongst developments in religion at the state level was the increasing emphasis upon the veneration of ancestor spirits, and the linking of these spirits with Hindu deities. During the Mataram period there appear to have been two classes of ancestor spirit: the primordial; and the proximate – those who had been prominent figures of the state during their lifetimes. The primordial ancestors were generally referred to as *sang hyang* (holy spirits).[8] They were apparently as powerful as the imported Hindu and Buddhist deities (Zoetmulder 1968:269), and some were clearly identified with mountains (Zoetmulder 1968:307ff).

The more proximate ancestors were normally addressed as *dewata*[9] and identified as *lumah* (lying) at specified sanctuaries. The relationship between the newly or historically deified dead (the *dewata*) and the major Hindu gods and goddesses (referred to as *bhaṭāra* and *bhaṭāri*) was apparently that of association. They often shared the same holy places, and these ancestors were venerated in conjunction with the Hindu gods – during the later ninth and tenth centuries mainly manifestations of Śiva. Ancestor kings played a

7 Wisseman Christie (1992:19–22); Degroot (2008).

8 *Hyang* is the general, indigenous term for the divine (Zoetmulder 1982:659).

9 *Dewata* is a term borrowed from Sanskrit, which was used in Old Javanese of this period both to denote some gods and also divine ancestors (Zoetmulder 1982:398). In Old Javanese inscriptions, however, the term was used solely to designate figures of importance in the state who had achieved apotheosis after death (Damais 1970:989).

central role in the 'myth' of the state that was constructed during this period, and appeals to these ancestors appear to have increased during periods of instability. With this in mind, a reconsideration of the inscriptions may provide insight into the geological conditions under which the state of Mataram evolved.

Reconsidering Mataram's Century of Turbulence

As noted above, the first clearly recorded change in Mataram's internal political climate occurred in 847 CE, when *rake* Pikatan took the throne. Although he was a member of the royal family of some importance,[10] his accession was not uncontested. Towards the end of his reign, having been forced to suppress an armed attempt to oust him, he moved the palace to Maḍang in Mamrati.[11] This move was apparently designed to deal with political unease: similar 'fresh-start' relocations are recorded in the histories of both this and subsequent Javanese states.

Rake Kayuwangi, who succeeded to the throne in 855 CE, both consolidated and expanded the state, taking control of the heartland of east Java by about 860 CE (Wisseman Christie 2001:42–45). His twenty-year reign – from which over sixty inscriptions have survived – was not, however, without internal strains. It is probably not a coincidence that the earliest surviving inscription from his reign, the metric Old Javanese Śiwagṛha inscription of 856 CE, was a eulogy dedicated to his recently deceased predecessor. The inscription summarizes the life of this 'Śaivite king', emphasizing his generosity as a builder of religious foundations, his successful defence of the state, and his establishment of a new palace at Maḍang. Since he had appeared prominently as a patron of the Buddhist temple Plaosan Lor in the 830s and 840s, he seems also to have changed religion.[12] In death *rake* Pikatan was linked to a Śiva-lingga, the heart-stone (*tĕas*) in an Abode of Śiva, at the centre of a temple complex equipped with platforms for ascetic practices and a sacred bathing place, where a tañjung tree was planted as a place of manifestation of holy [ancestor] spirits (*sang hyang*), in association with the *bhaṭāra*, the god Śiva (De Casparis 1956:280–330).

10 He appears prominently in the Plaosan Lor inscriptions of the 830s and 840s CE (De Casparis 1958).

11 See the Śiwagṛha inscription of 856 CE (De Casparis 1956:280–330; Wisseman Christie 2001:40–42).

12 See De Casparis (1958).

The Śiwagṛha inscription is a key document for the study of the formation of Mataram's state myth. Most of the major elements of the religio-political formulation are present in the text: the foundation of the palace of Maḍang by a guardian king, and his subsequent apotheosis as a deified ancestor, in a temple complex dedicated to Śiva, associated with asceticism, a sacred tree and holy water. This is the earliest surviving reference in Javanese inscriptions to a deified ancestor king. From this point on, however, they multiplied rapidly.

Although construction of his funerary temple may have begun during *rake* Pikatan's lifetime, it was completed by his successor, *rake* Kayuwangi. The temple of *rake* Pikatan's apotheosis has tentatively been identified as the Śiva temple in the major Hindu complex of Loro Jonggrang in Prambanan – presumably the first structure to be built in that large complex. This identification is given further weight by the contents of a series of six benefice charters, found in the area, that were issued by *rake* Kayuwangi and a close relation. This series, issued between 873 and 881 CE,[13] provided support for a temple complex dominated by a Śaivite tower temple (*prasāda*) called Gunung Hyang or 'Holy Mountain', to which an ascetics' hermitage (*kabikuan*) was attached. This seems to have been the last very large temple complex to have been built in central Java. The contents of the Śiwagṛha inscription suggest that it functioned as the state temple during the second half of the ninth century, linking a guardian ancestor king, in association with the god Śiva, to the welfare of the state.

Gunung Hyang was not, however, the only religious foundation to have been constructed and endowed with *rake* Kayuwangi's support. Overlapping in time with the endowments to Gunung Hyang, another series of endowments was undertaken by *rake* Kayuwangi and a second close relative. This second religious foundation appears to have been very different in nature from that of Gunung Hyang. The charters of this series[14] record support granted to a collection of about ten small Śaivite religious structures, including the religious house (*dharmma umah*) of Pastika, the sanctuary (*parhyangan*) of Smar, the hermitage (*patapān*) at Mulak, the religious foundation of Ramwi, and the tower-temples (*prāsāda*) of Upit, Kwak, and Laṇḍa, as well as those at Ruhur,

13 The Polengan series of inscriptions includes Tunhan, 873 CE; Humaṇḍing, 875 CE; Jurungan, 876 CE; Haliwangbang, 877 CE; Mamali, 878 CE; and Taragal, 881 CE (Suhadi and Soekarto 1986:2.7.1–2.7.6).

14 The Ngabean series and associated charters include Supit, 878 CE (Boechari 1986:52–53); Marsĕmu, c. 878 CE (Boechari 1986:53); Mulak I, 878 CE (Boechari 1986:28–30); Kwak I, 879 CE (Boechari 1986:30–32); Kwak II, 879 CE (Boechari 1986:32–33); Laṇḍa A and B, c. 879 CE (Boechari 1986:56–57); Ra Tawun I, 881 CE (Boechari 1986:37–38); Ra Tawun II, 881 CE (Sarkar 1971:xlix); Pastika, 881 CE (Sarkar 1971:li); Ra Mwi, 882 CE (Boechari 1986:38–41).

Tamwak, and Mataga. Some or all of these linked Śaivite holy places appear to have been located in Dieng, a region in the volcanic uplands of the Prahu massif also identified with Kailāśa, the abode of Śiva.[15]

At least two of these linked structures were clearly used as funerary temples. The tower-temple at Kwak had become, by around 882 CE, the *pacandyan*, or funerary temple, of the *rake* Wka, the member of the royal family who had co-sponsored the series of endowments with the king.[16] The religious house at Pastika was initially dedicated to the *bhaṭāra* (here apparently Śiva),[17] but after the king *rake* Kayuwangi himself died in 885 CE, the sanctuary precincts also became the locus of the 'holy *caitya* (funerary shrine) for the well-being of the *silunglung* of the *dewata* (deified ancestor) lying at Pastika'.[18] In the literature of Bali a small funerary pavilion called a *silunglung* is mentioned, built to contain the ashes of the dead, over which the priest says his prayers (Zoetmulder 1974:186). In addition, the benefice supporting the funerary temple at Kwak owed flower offerings to the funerary temple at Pastika,[19] thus preserving in death the hierarchical relations of the living.

The cohabitation – by Hindu gods and *dewata* – within the same sacred space is confirmed by another trend that is identifiable in charters issued after 855 CE. This was the increasing prominence of the curse formula in ceremonies for the dedication of benefices. Although earlier charters had occasionally threatened those disturbing a benefice with Hell and various unspecified punishments, imprecations had played a minor part in the records of early benefice ceremonies. But in 880 CE, towards the end of *rake* Kayuwangi's reign, there was at least one serous attack on the royal family. Part of the response, recorded in the benefice granted by the king in gratitude to those who had come to his family's aid, was a sharply increased emphasis on, and elaboration of, the curses pronounced against those who might undermine the throne. This first full curse text invoked the protection not only of the major and minor Hindu gods, but also of the 'celebrated ancestor spirits who protect the palace of the king of Mataram'.[20] This was the first, but not the last, invocation of deified ancestor kings for the protection of the throne. It was an appeal that was to become more frequent in the inscriptions of succeeding decades, as the state entered a period of growing turbulence.

15 See the Kapuhunan inscription, 878 CE (Sarkar 1971:xxxvii).

16 See the Laṇḍa A inscription, 882–883 CE (Boechari 1986:56–57).

17 For the Pastika inscription 881 CE, see Sarkar 1971:li.

18 Poh, 905 CE (Sarkar 1972:lxvi).

19 Inscription Kwak I, 879 CE. See note 14.

20 Wuatan Tija, 880 CE (Sarkar 1971:xlvi).

After *rake* Kayuwangi died early in 885 CE, three rulers succeeded to the throne within less than two years. The first two are said to have been driven out of the palace, and the third to have fled.[21] There followed a period of civil war that lasted for seven and a half years. It is possible that this political turmoil coincided with one of the periods of volcanic activity indicated by the ash layers at Borobudur.[22] Finally, late in 894 CE, the state was reunited by a ruler who immediately asserted his legitimacy by reconfirming a grant made by a deceased predecessor 'lying at Layang',[23] a place apparently associated with Dieng.[24]

Four years later, in 898 CE, *rake* Watukura *dyah* Balitung mounted the throne. Over forty inscriptions have survived from Balitung's twelve-year reign. The charters indicate that one of his first acts upon mounting the throne was to endow a series of holy places on the Dieng plateau and other volcanic massifs connected with royal ancestor spirits. He further consolidated his position by completing endowment projects of his deified predecessor, the *dewata* lying at Śataśṛngga, 'One Hundred Peaks'.[25] This holy site, named after the place where the ancestor of the Pāṇḍawas of the *Mahābhārata* died, appears along with Dieng in a list of mountainous abodes of spirits.[26] Balitung also added to the endowment of the funerary temple of *rake* Kayuwangi at Pastika,[27] thus linking himself to that successful pre-civil war ruler. In addition, he endowed a series of holy water places and other sanctuaries that were associated both with volcanoes and with ancestor shrines.[28] This continued the trend, begun by *rake* Kayuwangi, of associating ancestor kings with mountains.

By 905 CE, however, political unease appears to have returned. Balitung responded by moving the palace.[29] This political malaise may have been caused by increasing geological activity – earthquakes and minor eruptions – that culminated in a moderately serious volcanic eruption in central Java. In 907 CE Balitung was forced to create a new benefice and ancestor shrine (*kamūlan*) for

21 Wanua Tĕngah III, 908 CE (Djoko 1986; Wisseman Christie 2001:29–31).

22 Although no exact correlation can be proved, the Chinese ceramics found in the occupation layers between the narrow bands of ash appear to date to the late ninth and early tenth centuries (Mundardjito 1978, personal observation on field surveys 1977–99).

23 Panunggalan, 896 CE (Sarkar 1971:lviii).

24 Taji Gunung, 910 CE (Sarkar 1972:lxxx).

25 Tĕlang I and II, 904 CE (Suhadi and Soekarto 1986:2.6.1 and 2.6.2).

26 Kuṭi, c. 910 CE. The date equating to 840 CE in this garbled late copy is an error (Boechari 1986:16–21).

27 See the inscription of Poh, 905 CE (Sarkar 1972:lxvi).

28 See, for instance, Kubukubu, 905 CE (Boechari 1986:155–159).

29 Kubukubu, CE 905 (Boechari 1986:155–159); Wanua Tĕngah III (Djoko 1986; Wisseman Christie 2001: 29–31).

his grandmother in the area to the north of Magelang, because the previous benefice had been wiped out by such an eruption. The charter of the new benefice ends with an extended curse formula, reviving that used by *rake* Kayuwangi during a period of acute stress, calling not only upon the major and minor Hindu gods, but also the 'deified ancestors who protect the palace of the king of the state of Java'.[30]

That year was one of intense activity on Balitung's part. He endowed several other shrines associated with mountains, notably those on the slopes of the volcanoes Sumbing and Sundoro, devoted to the mountain spirits.[31] This particular charter ends with a curse formula that places even greater emphasis upon the ancestors. It drops reference to Hindu gods altogether, and instead calls upon indigenous spirits, including the 'holy ones of old (*hyang pūrwwa*), the purified servants, the floating fire – those guardian spirits, ancient ones, holy master spirits who have gone before, who held dominion over communities, lords of sanctuaries, builders of palaces ...those holy ancestors of Mḍang, of Poh Pitu, the great ones of Mataram: *sang ratu* Sañjaya, *śrī mahārāja rakai* Panangkaran, ...Panunggalan, ...Warak, ...Garung, ...Pikatan, ...Kayuwangi, ...Watu Humalang...'. This list includes only those earlier kings judged to have been successes. The failures were not accorded guardian ancestor status, even if they achieved apotheosis after death.

The sense of unease clearly continued in the state after the eruption of 907 CE. In the following year Balitung was advised to revive the benefices of all the Buddhist foundations of the state in order to secure the stability of the throne.[32] This approach seems not to have produced the desired results, since charters issued in 910 CE returned to appeals to ancestor and mountain spirits. One of these called upon spirits inside Mt. Wilis, associated with the 'holy Fire' (*hyang api*), for protection.[33] And the Kuṭi inscription of the same period ends with a curse calling upon Śiva and the spirits of holy places of volcanoes, including Dieng, Śataśṛngga, Merapi, Sumbing and Sundoro, along with the holy spirits of Mḍang.[34] By this time the ancestral palace site of Mḍang/ Maḍang had become famous enough for its name to have been given to a place in Luzon in the northern Philippines, at the farthest reaches of Java's sphere of influence.[35]

30 Rukam, 907 CE (Nastiti et al. 1982:23–28).

31 Mantyāsih I, 907 CE (Sarkar 1972:lxx).

32 Wanua Tĕngah III, 908 CE (Djoko 1986; Wisseman Christie 2001:29–31).

33 Pĕnampihan, 910 CE (Sarkar 1971:lix).

34 Kuṭi, c. 910 CE (Boechari 1986:16–21).

35 Laguna, 900 CE (Postma 1991).

Although *rake* Hino *pu* Dakṣa succeeded to the throne uneventfully in 910 CE, he clearly inherited a troubled state. He, too, appealed to the ancestors. While Balitung had emphasized the link between guardian ancestors and volcanoes, Dakṣa focused more exclusively upon the ancestor kings themselves – 'those who formerly ruled the world'.[36] Like Balitung, he honored *rake* Kayuwangi,[37] and he also venerated Balitung.[38] More importantly, however, he brought to the fore the early eighth-century founder of Mataram, *sang ratu* Sañjaya. Not only did he grant a benefice to the lineage temple of the descendants of Sañjaya,[39] he went so far as to date four of his twelve surviving inscriptions in the Sañjaya era[40] – the first break in Mataram's history with the otherwise universally used Indian Śaka era.

Dakṣa's use of the Sañjaya era suggests an attempt to maintain political unity through reinforcing connections with a legitimizing past – a step beyond appealing to guardian ancestors of the state. This move appears to reflect the deliberate construction of a foundation myth for the state focused upon a founder-ancestor. Although none of his successors followed his rather extreme example, ancestors remained a major theme in the inscriptions of succeeding kings.

Rake Layang *dyah* Tlodhong mounted to the throne in 919 CE – again apparently unopposed. Yet he, too, felt the need to link himself to past rulers. His first recorded act was to endow his father's funerary shrine,[41] and the next, to settle a dispute concerning a grant made by the revered former ruler *rake* Kayuwangi.[42] In fact, almost all of his charters involved shrines connected with ancestor figures, who were described in one charter as 'those who were *made* holy ancestors of old',[43] highlighting the apotheosis of the dead. Most of these ancestors were connected with the Śaivite sanctuaries of Dieng or of other volcanic uplands of central and east Java.

Tlodhong's inscriptions returned to the nagging problem of geological instability. His curses called upon not only Hindu gods and royal ancestor spirits, but also the *nāgarāja*, the serpent kings, 'who in the depths make firm the sacred mountains'.[44] In Hindu mythology, these *nāgarāja* were inhabitants of

36 Taji Gunung, 910 CE (Sarkar 1972:lxxx).

37 Wintang Mas A, 914 CE (Sarkar 1972:lxxxviii).

38 Taji Gunung, 910 CE (Sarkar 1972:lxxx).

39 Pupus, c. 910 CE (Boechari 1986:75–77).

40 Taji Gunung, 910 CE (Sarkar 1972:lxxx); Timbanan Wungkal, 913 CE (Brandes 1913:xxxv); Tihang, 914 CE (Boechari 1985); Tulang Er, 914 CE (Santosa 1994).

41 Lintakan, 919 CE (Sarkar 1972:lxxxvi).

42 Wintang Mas B, 919 CE (Sarkar 1972:lxxxviii).

43 Gilikan I, c. 923 CE (Sarkar 1972:civ).

44 See above.

the caves of inaccessible mountains: when the *nāgarāja* yawns, the earth trembles, and when he vomits fire he destroys (Daniélou 1964:308, 163). Wrapped around mount Mandara (or Meru), a *nāgarāja* provided the means by which the gods and demons churned the ocean of milk to produce *amṛta*, the elixir of life (Zoetmulder 1968:308–309). These Hindu serpent kings, associated both with the stability and the agitation of mountains, were absorbed into the array of spirits governing the stability of Java's mountains. One manifestation of this appropriation is the central figure in the tenth century water temple Jalatuṇḍa on the slopes of Mt. Penanggungan in east Java – a water spout in the form of a nine-peaked lingga-mountain encircled by a *nāgarāja* (Bosch 1961:52).

Geological instability may explain why the next ruler, who came to the throne in around 927 CE, was displaced before the end of that year. The first act of his successor, *rake* Sumba *dyah* Wawa, was to reconfirm a grant made by his recently deceased predecessor, described as 'lying at Kayu Ramya', but not accorded the apotheosis title of *dewata*,[45] perhaps reflecting his failure as a guardian figure. Wawa then went on to have a Śivamaṇḍala constructed to 'gain recognition by the holy spirits (*hyang*)'.[46] Such constructions, performed in connection with ceremonies for the dead, are mentioned in the *Agastyaparwwa* (Gonda 1933–36:336.16, 380.7). The early tenth-century Śaivite version of the *Sang Hyang Kamahāyānikan* also lists the Śivamaṇḍala amongst the vehicles for the attainment of heaven (Chandra 1997:19, 6a). It seems likely that this particular ceremony was held to promote the apotheosis of Wawa's own father, mentioned at the beginning of the charter as 'lying at Alas', but not yet bearing the title of *dewata*.

These two inscriptions were apparently the last to be issued in central Java. Although Wawa issued at least eight other charters during his two-year reign, all of these were in east Java. This burst of grant issuing, and the shift of geographical focus to east Java, may reflect not only the lure of the increasingly prosperous east Javanese ports, but also instability and uncertainty in central Java. Curse formulas again became prominent in the charters, even those issued in east Java – another sign of unease. These imprecations called upon both Śiva and ancestor spirits – 'those who are known to protect the palace of the king of Mḍang in the state of Mataram'.[47] Wawa's charters that placed the most emphasis upon the guardians of the ancestral palace of Mḍang were those farthest removed from its original site. They may have been paving the way for a shift of the palace to the east as the situation deteriorated in central Java.

45 Wangwang Bangĕn, 927 CE (Boechari 1986:14–16).

46 Wulakan, 928 CE (Sarkar 1972:xciv).

47 Sangguran, 928 CE (Sarkar 1972:xcvi); Kampak, 928–929 CE (Brandes 1913:lii).

This period of growing unease came to a head in 929 CE. Wawa was succeeded by his heir-apparent, *rake* Hino *pu* Siṇḍok. Although more than twenty-five inscriptions survive from Siṇḍok's twenty-year reign, none was issued in central Java. In fact, it appears that no inscriptions were issued in central Java by the state of Mataram after 928 CE. A catastrophe of some sort must have occurred there. Most of the officials listed in Wawa's administration also appear in the early inscriptions issued by Siṇḍok in east Java, suggesting that political breakdown was not the problem. And no other power is recorded as having moved into the heartland of central Java to replace the court following its shift to east Java. Since, as noted above, low-lying sites of this period lie under several layers of volcanic mud, it seems most likely that this catastrophe was caused by a major eruption – or series of eruptions – of Merapi, the volcano closest to the old heart of the state.

Siṇḍok began his reign in east Java by issuing a rapid burst of benefice grants to religious foundations, apparently in an effort to establish authority and stability. The eight surviving charters issued during his first year on the throne were focused upon ancestor shrines linked to Śaivite gods, holy water foundations and mountains.[48] The curse formulas of these charters called upon the protection not only of Śiva and the deified ancestors who 'protect the palace of the king of the state of Mataram', but also of the 'Nāgarājas, the serpent kings, whose encircling coils stabilize the sacred mountains'.[49]

Siṇḍok again moved the palace, to Tamwlang, in 929 CE,[50] possibly in an attempt to escape bad luck in a time of political turmoil. He seems not to have succeeded, since two years later he was forced not only to reward clients who had provided military aid to secure his seat on the throne, but also to attempt to enhance his image by presenting himself as 'the manifestation of the Sun embodied as the god Śivāditya'.[51] Despite his unprecedented assumption of deified status while still alive, he appears to have failed to stabilize the state, since in 933 CE he was still rewarding clients for military aid.[52]

Consequently, by 934 CE, Siṇḍok's charters began to direct resources towards a series of religious establishments that were, as before, lineage temples focused upon ancestors of the royal family, but now no longer purely Śaivite in nature. Śiva and Buddha were merged in several foundations – some new, some

48 These include Gulunggulung, 929 CE (Nakada 1990:2–21); Linggasuntan, 929 CE (Nakada 1990:21–33); Cunggrang, 929 CE (Stutterheim 1925); Jĕrujĕru, 930 CE (Nakada 1990:33–51).

49 Turyyan, 929 CE (De Casparis 1988).

50 See above.

51 Waharu IV, 931 CE (Boechari 1986:59–65).

52 Gĕwĕg, 933 CE (Brandes 1913:xlv).

converted to this new syncretic mode.[53] This increasing syncretism was accompanied by an emphasis upon tantric knowledge and asceticism, as expressed in the Buddhist/Śaivite treatise *Sang Hyang Kamahāyānikan*, attributed to his reign.[54] The imprecation formulas of the charters continued to call upon the guardian ancestor kings. In addition, at least two of these foundations were focused upon *amṛta*, the elixir of life,[55] and charged directly with praying for the security of the throne. These endowments paralleled those of temples linked to the holy water temple at Walaṇḍit, where the focus of worship was the major volcano Mt. Bromo.[56] Despite these attempts to find a formula that would provide stability, Siṇḍok was apparently forced to move the palace yet again – to Watu Galuh – by 943 CE,[57] where he appears at last to have achieved some sort of security for himself and his successors, since the palace settled there for the next 40 years.[58]

Siṇḍok's reign, which ended in about 948 CE, was apparently a difficult one. The loss of the old central Javanese heartland clearly cast its shadow over its early years. The fact that the state remained troubled through most of his reign may reflect not only the profundity of that loss, but continuing geological turbulence. The dearth of charters issued by his immediate successors and the stable site of his final palace, however, suggest that a degree of internal peace had returned to both the state and its environment.

This apparent political stability lasted for little over half a century, until the state suffered the next period of misfortune. By the beginning of the eleventh century, the state was yet again under attack – this time, in 1016 CE, by forces led by *haji* Wurawari of Lwarām, who burned the palace and killed the king and a large portion of the court. The record of this disastrous attack, and its aftermath, is contained in the Pucangan charter, issued in 1041 CE by the dead king's successor Airlangga[59] – who accords his deceased predecessor the status of *dewata*, despite his defeat. Although Airlangga managed to reassemble the state, it broke down yet again in civil war following his death in about 1048 CE, and remained divided until the thirteenth century. Whether geological instability

53 Hĕring, 934 CE (Brandes 1913:xlvii).

54 See Kats (1910) for the Buddhist version; Chandra (1997) for the Śaivite version.

55 Añjukladang, 937 CE (Brandes 1913:xlvi); Sobhāmṛta, 939 CE (Cohen Stuart 1875:xxii, plates I and 3: for the other 4 plates, unpublished ms by Wisseman Christie 2003 and Titi Surti Nastiti 2004).

56 See Muñcang, 944 CE (Brandes 1913:li).

57 Paradah II, 943 CE (Brandes 1913:xlviii).

58 See Bandar Alim I, 985 CE (Suhadi and Richadiana 1997:2.10.1).

59 See Kern (1885); Brandes (1913:lxii); Poerbatjaraka (1941); for revised date, see Damais (1952:64), n. 2.

played a role in Mataram's final breakdown is uncertain, although the east Javanese water temple of Walaṇḍit – apparently dedicated to an ancestor figure (*dewata*) and associated with the active volcano Bromo – first mentioned in a charter of Siṇḍok,[60] continued to be a focus of veneration in succeeding centuries.[61]

Conclusion

Power politics in early Java were as turbulent as they were in many places at the time, but they were played out in a physical environment that was both more fertile and more unstable than most. Volcanic activity, in particular, has influenced the politics of the island for centuries. The contents of inscriptions surviving from the late first and early second millennia CE cast an oblique, but useful, light on the effect of environmental stress upon one of the island's early states – that of Mataram – and of the uses made of religion in generating and sustaining political stability.

A state, to survive, requires the security not only of stable institutions, but also of unifying ritual and myth – and Mataram survived for over 300 years, at times in the face of considerable odds. Of the three centuries of Mataram's existence, it was during the years between 847 and 948 CE that the state's resilience was most severely tested. Yet, although it apparently abandoned its original core region in central Java during the second quarter of the tenth century, the state continued to exist into the middle of the eleventh century (Wisseman Christie 2001). Examined against the background of apparent geological instability, certain trends in the charters issued by Mataram's rulers between 847 and 948 CE become significant. By the end of Siṇḍok's reign in 948 CE, the process – begun under *rake* Kayuwangi a century before – of merging imported Buddhist and Hindu gods with indigenous ancestor spirits had produced a syncretic formulation that was still recognizable centuries later. The upheavals of the ninth and tenth centuries also produced a state foundation myth and ideology linked to venerated ancestors and volcanoes that persisted even longer. The institutions that Mataram had formed in good times remained durable, and in bad times – even during the state's catastrophic truncation – its religious and ideological framework proved to be adaptable.

60 See Lingasuntan, 929 CE (Brandes 1913:xxxix).

61 See the Walaṇḍit charter of 1405 CE (Brandes 1899:64–69).

CHAPTER 5

History and Seismology in the Ring of Fire

Punctuating the Indonesian Past

Anthony Reid

Terrifying spasms of the planet like the one that struck northern Honshu in March 2011, or Aceh and much of the Indian Ocean littoral on Boxing Day 2004, force us to reflect on what our discipline is about. The longer the time-depth considered, the more human history can be seen to be dependent on the beneficence of the planet we inhabit. In our preoccupation with the rise, fall, successes and follies of states, especially of our own states, historians have often missed the larger vocation of seeking to understand the planet itself. Here Peter Boomgaard blazed a crucial trail, showing what appropriately equipped historians can do to write the history of the Southeast Asian environment as a crucial part of that of the planet.[1] Yet this too risks becoming a marginal sub-discipline. The great majority of historians of Southeast Asia focus on the nineteenth and twentieth centuries, and have little contact with those we call pre-historians, let alone seismologists, geologists, life scientists and climatologists.

Past Neglect

Since 2004 Sumatra has come to be understood as one of the most active and dangerous tectonic subduction zones in the world, with the most dramatically evident periodicity of mega-events (Perkins 2008). But the historians and others who wrote about Sumatra in the late nineteenth and twentieth centuries did little to prepare their readers for violent earthquakes, tsunamis, volcanic eruptions and climatic aberrations. Authors such as C. Snouck Hurgronje, E.S. de Klerck, E.B. Kielstra, John Bastin, Denys Lombard, James Gould, Mohammad Said, Ibrahim Alfian, Christine Dobbin, J. Kathirithamby-Wells, Freek Colombijn and myself virtually ignored natural events, though we have done a little better since 2004.[2]

1 Firstly by leading the KITLV project on Ecology, Demography and Economy in Nusantara (EDEN) for 15 years, and finally in his path-breaking *Southeast Asia: An Environmental History* (Santa Barbara: ABC Clio, 2007).

2 To be fair, Freek Colombijn (1994:43) had one sentence noting that Padang was virtually destroyed by the 1797 earthquake. Since 2004 we have all done better, including Colombijn in

 | DOI 10.1163/9789004288058_006

The disasters which have succeeded one another between the Aceh-Nias tsunami of December 2004 and the South Pagai (Mentawai) tsunami of October 2010 (Table 5.1) appeared unprecedentedly ferocious to Indonesians living today, but were 'business as usual for the earth' (McCaffrey 2009:363). We now know that such ruptures in the earth's crust occur with necessary regularity, and that the bigger events every few centuries have enormous impacts not only on Indonesia but on the planet as a whole. Our politically-focussed histories look woefully inadequate to explain this pattern.

The regularity of disasters on the subduction zones was reported first for California and Japan, while Sumatra was little noticed even by geologists until a seminal article by geophysicists Newcomb and McCann (1987) assembled the evidence pointing to its dangers in 1987. The scientists use the data of historians when they can, though in Southeast Asia we have not been particularly helpful to them. A comprehensive record began to be available for the whole planet only once the seismograph was invented around 1900, and long after that descriptions of the effects of seismic movements on human

TABLE 5.1 *Major seismic disasters in Indonesia since 2004*[3]

Date	Magnitude	Place	Direct Fatalities
2004, 26-12	9.2 M + tsunami	Aceh-Nias	167,800
2005, 28-3	8.6 M + tsunami	Nias	1,313
2006, 26-5	6.3 M	Yogyakarta	5,749
2006, 17-7	7.6 M + tsunami	Pangandaran, W. Java	802
2007, 12-9	8.4 M	Mentawai, Padang,	25+
2009, 30-9	7.6 M	Padang	1,117
2010, 25-10	7.7 M + tsunami	Mentawai	400
2010, 26-10	V4 eruption	Merapi, Java	324
2012, 11-4	8.6 earthquake	Aceh	5
2013, 15-9	V2 eruption	Sinabung, N. Sumatra	17
2014, 13-2	V2 eruption	Kelut, E. Java	7

several articles, and Jeffrey Hadler (2008: ch. 6). My edited *Verandah of Violence: the Background to the Aceh Problem* (Singapore University Press: 2006) was named partly in recognition of Aceh's exposure to tectonic violence.

3 Earthquakes are rated by magnitude (M), volcanic eruptions by Volcanic Explosivity Index (V) on a scale from 1 to a maximum 7, following the Smithsonian Institute Global Volcanism Program.

societies remained underreported in Indonesia. Before 1900 our understanding is dependent firstly on the written records of observers, and increasingly on scientific investigation of the traces left by volcanic eruptions, earthquakes and tsunamis on the ground. The ring of fire around the Pacific where the periodic pattern should be clearest is for the most part deficient in historical records before 1600, perhaps in part because of the pattern of periodic disasters. Only Japan has records back into the first millennium CE. These have preserved a pattern of earthquakes with attendant tsunamis in the Nankai subduction zone off Shikoku and southern Honshu as far back as 684, with the biggest ones in 1293 (Kamakura – over 20,000 estimated killed), 1703 (Genroku, over 100,000 perhaps killed, est. 8.0 M), in 1605 (est. 7.9 M), 1707 (8.4 M), 1854 (8.6 M) and 1946 (8.4 M). Another huge one is therefore anticipated in that region anytime, although it was not anticipated further north in Tohuku.

Our lack of knowledge of Indonesia's tectonic past, in contrast not only with Japan but with other less populated or exposed places around the ring of fire, is strikingly borne out in the most influential data sets now available to the scholarly community. The US National Geophysical Data Center records only 26 tsunamis known in Indonesia before 1800, of which 20 were in the areas of immediate concern to the lucrative VOC spice monopoly in Maluku in the seventeenth and eighteenth centuries. Four large ones were reported at the European pepper-posts of Padang and Bengkulu in Sumatra, but none in Java where the endangered south coast was then of no interest to the Dutch. By contrast 29 have been recorded in the 22 years since 1990, including 9 massive ones killing over 100 people, two of them in Java.[4] Similarly for volcanic eruptions, the Smithsonian Institute list shows that since 1982 there have been 25 eruptions in the world of a scale to cause great damage (4 or more on the Volcanic Explosivity Index, or VEI, a scale extending to 7), of which 10, or 40%, have been in Southeast Asia or Papua New Guinea, 5 of them (20%) in Indonesia. Most of the others have been in underpopulated areas of the ring of fire such as Alaska or Kamchatka. The further we go back in time, however, the less well represented Indonesia is in the record, with 16% in the nineteenth century, 3% (just one recorded) in the eighteenth, 12.5% in the seventeenth, 9.5% in the sixteenth, and nothing at all in the acknowledged record before that except a suspect memory of an eruption of Krakatau in 416 AD. By contrast hundreds of eruptions over the last few thousand years have been scientifically dated elsewhere, notably in

4 NOAA National Geophysical Data Center, register of tsunami events, http://www.ngdc.noaa.gov/hazard/hazards.shtml, accessed 6 July, 2011.

northern sections of the Pacific Ring of Fire and in New Zealand. Almost everything remains to be done in Indonesia.[5]

Why has there not been more attention to the earthquakes, tsunamis and volcanic eruptions that punctuate Indonesia's history? Indonesia's relative poverty and weak scientific infrastructure is of course the major reason. But in addition, the twentieth century in which adequate reporting really began appears to have been exceptionally quiet seismologically, while it was full of violence and upheaval politically. In the nineteenth century Indonesia's massive volcanic eruptions, Tambora (1815, VEI 7) and Krakatau (1883, VEI 6), darkened and cooled the whole planet and produced years without summer in Europe, though this causation has only recently been understood (Zeilinga de Boer and Sanders 2002: ch.6, ch.7). In the twentieth there were the usual frequent eruptions, including eight which have subsequently been ranked as VEI 4 for their destructive capacity. Only Gunung Agung's 1963 eruption in Bali was rated a 5 and killed some 2000 people. Even more remarkably, no mega-ruptures were recorded on the subduction zones of Sumatra or Java causing earthquakes in excess of magnitude 8 between the 1861 one in Southwest Sumatra (later reconstructed as 8.5 M) and the massive quake (9.2 M) and tsunami of 2004. Further east, earthquakes of 8.5 and 8 M were reported in Maluku in 1938 and 1939, an 8 in Sumba and Sumbawa in 1977, and an 8.2 in Papua in 1996, but these were in remote and less populous areas that made little impact on urban Indonesians.

Thirdly, before twentieth century changes in urbanization and building styles, Indonesians themselves were little affected by earthquakes, and appear to have avoided settling in large numbers on the coasts of Nias, western Sumatra and southern Java which were exposed to tsunamis. The people of Nias, probably the most vulnerable to tsunamis of all complex Indonesian societies, spurned their coasts completely before sea-based Dutch infrastructure arrived in the second half of the nineteenth century. They fished only in fresh-water rivers, and built their villages on hilltops where possible (Gruber 2007). Mentawai islanders were also extremely tsunami-prone and built their villages well away from the coast (Loeb 1935:134, 161). Moreover, as

5 Smithsonian Institute Global Vulcanism Program, 'Large Holocene Eruptions', http://www.volcano.si.edu/world/largeeruptions.cfm, accessed 7 June, 2012. Peter Boomgaard's list culled from Dutch sources has 13 eruptions between 1663 and 1696, all in Maluku thanks to the concentration of Dutch interest there, except for one in Flores. More research is needed to give a VEI rating to any of these, all different from the four large ones (VEI 4 or above) listed by the Smithsonian (Boomgaard 2001b:212).

noted by Marsden ([1811] 1966:30) in the late eighteenth century, 'the usual manner of building [in light wood, bamboo and thatch], renders them [earthquakes] but little formidable to the natives'. In the more geologically active period of the eighteenth and early nineteenth centuries European settlers did treat earthquakes as a major problem for their buildings. Marsden was one such, while Alexander Hamilton ([1727] 1930, II:61) reported of Bengkulu in the early 1700s that 'Brick or Stone Walls in that Country cannot long continue firm, because Concussions of the Earth are so frequent by Earthquakes, that solid walls are rent by the shaking of their Foundations'.

The Malay historical and literary texts most accessible to scholars have been of little help in reconstructing the record. Ian Proudfoot's splendid Malay Concordance enables one to track the usage of terms relating to earthquakes in the hundreds of texts now accessible electronically. *Gempa bumi* (earthquake) in some form occurs indeed 34 times on that site, but 27 of the instances are from the Malay Bible and the remaining seven (mainly from Raniri's encyclopedic *Bustan as-Salatin,* written in Aceh in the 1640s but often oblivious to local conditions), also show the wrath of God through earthquakes in the central lands of the Abrahamic tradition, not in Sumatra or Java.[6] The much greater seismic mega-events of the Indonesian subduction zones appear not to have become as central to the classic Malay texts as rare events in the holy lands of the Middle East.

Only since 2004 have scholars begun looking in other corners of oral and written Indonesian literature for more clues about past seismic events. Volcanic eruptions are more frequently noted in the chronicles of Java and Bali, though as omens rather than a source of destructive change. Although seismologists have hoped the Java specialists would find confirmation in them for the mega-eruptions that darkened the planet in 1257 and 535 CE, the science has so far been more useful than the chronicles.[7] Oman Fathurahman recently found scribbled (in Arabic) on the front page of an old Arabic manuscript by al-Ansari preserved in the religious school of Tanoh Abee near Banda Aceh the

6 http://mcp.anu.edu.au, accessed in April 2010. In addition to variations of *gempa* and *gempa bumi,* I did a search for *lindu,* a possibly older Javanese-derived term. This yielded 102 occurrences, all in a single *Kitab Takbir* (Book of Omens). Drawing on some venerable Islamic traditions as well as pre-Islamic local ones, this used earthquakes as one of the many natural phenomena to be exploited for an understanding of favourable and unfavourable times for action.

7 David Keys (2000) made speculative use of the 19th century *Pustaka Raja Purwa* to point to Krakatau as the source of the 535–6 planetary cooling. An interdisciplinary group recently found more convincing evidence for a mega-eruption in Lombok in 1257, though reconciling the chronicle evidence with this is a task urgently awaiting attention (Lavigne et al. 2013).

comment that 'a second destructive earthquake' occurred on Thursday 9 Jumadil akhir 1248 H (3 November 1832) (Fathurahman 2012). This suggests there were major Aceh earthquakes leading up to the well-documented earthquake and tsunami which destroyed many buildings in Padang and Bengkulu on 27 November 1833.

An Acehnese scholar has collected four related manuscripts, hitherto ignored, devoted to the concept of *tabir gempa*, or the interpretation of earthquakes. The concept *tabir* is more often applied to dreams, implying that hidden meanings can also be drawn from these spasms of the earth. Two of the manuscripts were located in Banda Aceh, one in Padang and one in the National Library collection in Jakarta, but all appeared to derive from the same traumatic experience of the earthquake and tsunami which caused much damage to Padang, Bengkulu and presumably other parts of Sumatra's west coast in November 1833. Variously dated from the evening of 24 November until the 27th, this is the only major earthquake and tsunami of the late eighteenth or nineteenth century to occur in the Muslim month of Rajab (1249 AH),[8] and all four manuscripts stress the significance of that month. One declares 'An earthquake that occurs in the month of Rajab, should it happen at *subuh* (first light) then the country will suffer, should it happen at *dhuha* (sunrise) then the water from the sea will hit hard'.[9] We may at least assume that the worst impact of the 1833 tsunami hit Aceh's west coast at dawn, probably on 25 November.

Clues to earlier events in the indigenous literature are even more sparse. The Hikayat Aceh says its hero (later Sultan Iskandar Muda) 'felt the earth move like the sensation of an earthquake', making his horse also tremble as an omen of coming events – in the early 1600s (Iskandar 1958:156). In Barus, north of Sibolga, several different chronicle traditions tell the story of how Barus was recognized as independent of Aceh, perhaps also in the seventeenth century. In this story Sutan or Sultan Ibrahim of Barus was killed and beheaded by soldiers of the Sultan of Aceh. When the victorious sultan kicked the head of his vanquished rival in scorn, however, he was punished by God with a terrible

---

8 The following month, Shaban, would surely have been stressed by any later manuscripts, since both a moderate 1836 earthquake and tsunami, and the massive one of 1861, occurred in that month (as had the 1797 quake and tsunami that destroyed much of Padang).

9 '*Jika pada bulan Rajab gempa ketika Shubuh alamat isi negeri itu sakit padanya. Jika pada ketika Dhuha air laut akan keras padanya*', cited in Hermansyah, 'The Acehnese Adaptation and Mitigation on Earthquake and Tsunami in the 18th Century (An Ancient Manuscript Research)', paper prepared for 3rd ICAIOS Conference, Banda Aceh, May 2011, kindly made available by the author.

disease, and 'a flood devastated the country, carried away rice and cattle, and decimated the population'. Hence the Acehnese repented, and swore that the freedom of Barus would be forever respected by Aceh (Chambert-Loir 2009:512; Drakard 1990:100,102).

A Relatively Quiet 120 years

Tectonic movements are predictable, resulting from the relentless movement of the oceanic plates 'subducting' under continental plates as they collide, often in deep trenches. In the Sumatran case, the Indian Ocean plate is advancing at about 65 millimetres a year, in the middle range of those around the world, a pressure that must be periodically released in earthquakes (Heidarzadeh et al. 2008:776). But until the current round of disasters in Sumatra and Java since 2004, neither historians nor scientists had been jolted into giving Indonesia's past record the attention it deserves. Indonesia had, it now appears, been unusually devoid of major natural disasters for 120 tears after a turbulent period in the nineteenth century which had included the eruption of Tambora in 1815. This was the biggest volcanic disaster anywhere of which history has records, assigned the maximum VEI of 7, its ash emissions blamed for producing a year without summer in much of the northern hemisphere.[10] The same century witnessed the Krakatau eruption in 1883 (VEI 6), the world's second biggest in modern times, which created a dust veil carried twice around the world, which lowered global temperatures significantly. If such an event occurred tomorrow it would ground airplanes throughout the world, and create chaos for all of us. But Krakatau occurred before the age of modern measurement and record-keeping, and before the airplane.

In the geologically calmer period between 1885 and 2000 the Indonesian population grew from about 25 million to 225 million, and its urban (predominantly coastal) population from little over a million to 129 million. The period of modern measurement of earthquakes through the seismograph (from 1900) coincided with a period of misleading geological moderation, so that modern Indonesia could be built without serious consideration of the dangers it would

10 The largest volcanic threat to *homo sapiens* thus far appears also to have arisen in Indonesia, in the biggest eruption of them all 74,000 years ago which created Lake Toba in Sumatra. Peter Boomgaard (2007b:31) points out that this darkened the planet for six years and reduced mankind to a few thousand – 'the last almost successful attempt by "nature" to rid itself of this peculiar brand of primate'.

eventually have to face. Since 1870 the importance of colonial ports as hubs for each province has pushed the population to cluster dangerously on this narrow plain. In a pre-2004 article I estimated the population in 1830 of the coastal plain of what is now the Province of West Sumatra had been only one tenth that of the uplands of the same province. In the Batak areas further north (present North Sumatra Province), the contrast was even greater, with about 40 times the population in the highlands as along the western coast (Reid 1998a:61–89).

In the interval between Padang's earthquakes of 1833 and 2009, the population of the West Sumatran littoral had soared from 80,000 to over two million, and for the first time surpassed the population of its safer and healthier highland hinterland. Moreover, a high proportion now lives in sub-standard but 'modern' brick and concrete dwellings highly subject to earthquake damage, whereas in past events only the handful of Europeans were affected by the collapse of such buildings. Hence the earthquake of September 2009 centred off the coast south of Padang caused a death toll of over 1000 with 135,000 homes destroyed and 1.25 million people affected, although only registered as magnitude 7.6. The much bigger shocks of 1833 and 1861, by contrast, had caused at most a handful of deaths. In Java, similarly, the exposed southern coast was almost unpopulated until the nineteenth century, but the Dutch foundation of a major base at Cilacap in 1840, and the subsequent location of an oil refinery there, have seen the population of the southern littoral grow to about two million today.

The relatively low number of casualties from Indonesia's natural disasters in the twentieth century, and the very high numbers of casualties of political conflict, mostly state-sponsored, make it unsurprising that historians should have focused on the political. Table 5.2 shows that the deaths directly attributed to political conflict in the twentieth century outnumbered those caused by geological spasms by 30 to 1. The reversal has been dramatic since 2000, chiefly because of the extraordinary death toll of the 2004 tsunami. Natural disasters outnumbered conflict deaths 17 to 1 in the first decade of the present century. Steven Pinker has recently shown that violent deaths from human conflict have in fact declined dramatically in most places as a proportion of total deaths (Pinker 2011), even if the Indonesian figures look high in the mid-twentieth century as the world-system readjusted from empires to nation-states. Since 1980 Asia has joined the global downward trend for violent conflict deaths, which looks like being permanent. Natural disasters on the other hand have already caused more deaths in Indonesia in the first decade of the twenty-first than the whole of the twentieth century, and will certainly cause many more.

TABLE 5.2 *Deaths from natural disasters and human conflict in Indonesia*

	Earthquake deaths[11]	Tsunami deaths	Volcano direct deaths	Total natural disasters	Approx. direct Conflict deaths
1900–09		400	76	476	21,000
1910–19	15,020		5,100	20,120	2,500
1920–29	777	128	263	1,065	450
1930–39	23		1,379	1,402	50
1940–49	213		117	330	363,000
1950–59	8		71	79	40,000[12]
1960–69	236	264	1,766	2,077	500,000
1970–79	1,410	175	727	2,235	65,000[13]
1980–89	880	11	268	1,145	5,000
1990–99	3,040	223	118	3,269	10,200
Total C20th	21,607	1,201	9,665	32,473	1,028,000
2000–09	8,324	168,300	2	176,626	10,700

Scientific Evidence Now Extends the Record into the Past

The seismologists have of course taken full advantage of historically known dates of earthquakes and tsunamis to help them plot the periodicity with which they occur in different zones. But science has also developed techniques of dating these seismic events at periods long before any historical record, though without, of course, the same exactness about dates or human consequences. The results from Japan have again led the way, notably for the subduction zone of Hokkaido, one of the more dramatic points at which the Pacific plate meets the Japanese island chain. A recent study of sand deposits

11 These first two columns (earthquake and tsunami deaths) are taken from Asian Disaster Reduction Center (*Data Book* 2000).

12 Micheal Clodfelder (1992, II:1127) gives 33,414 killed in PRRI/Permesta conflict of 1958–1961, most in 1958. The remainder of the total is a conservative estimate of conflict deaths in the Darul Islam rebellions in Aceh, West Java and Sulawesi.

13 Clodfelder (1992, II:1140) estimated 60,000 East Timorese killed by Indonesian troops in 1975–1984, 2,000 on the TNI side in 1975–1976 alone, and 2,000 in the East Timor Civil conflict of 1975.

in a lagoon and adjacent beach-ridge plain near Kiritappu provided evidence of 15 tsunamis over the past 6000 years, only the last two of which, in 1843 and 1894, are known to history in this frontier for Japanese settlement. Approximate dates were established by carbon-dating of the detritus in the lagoon above and below each deposit layer. By this means it was calculated that the mega-tsunamis occurred at an average of 400 years interval, though varying unhelpfully from 100 to 800 years. Most chilling was the conclusion, relevant also for Sumatra, that 'most of these tsunamis were larger than any generated at Hokkaido in the last 200 years' (Yuki Sawai et al. 2009).

Although very little comparable work was done on Sumatra before the 2004 disaster, the island has now been declared by a Scripps Institution geodesist to be 'the best area in the world to be able to predict a quake', because of the perceived periodicity of major tears in its subduction zone (Yehuda Bock, as cited in Perkins 2008). The work of Newcomb and McCann (1987:43–44) well before the 2004 tsunami trauma had already established the 1861 earthquake as the most recent rupture of the northern zone of the Sumatra subduction plate, and the 1833 quake as having the same role in the South-Central section. The more abundant research since 2004 appears to have accepted these two events as the most recent critical ones, but sought to push the frontier of knowledge further back in time through physical investigation on the ground.

One method is to examine the layer of sand deposited by a tsunami in the 'swales', or depressions between beach ridges, where cases can be found that are relatively undisturbed by subsequent events. Dating is provided by carbon-dating the organic matter accumulated immediately before and after (i.e. below and above the layer), for an approximate time depth. By these means two international teams of researchers were able to show several layers of sand deposit between the organic accumulated material both at the small tsunami-affected island of Phra Thong, north of Phuket, and on the west coast of Aceh a little north of Meulaboh. At both sites there were older tsunami events, of which the one dated around 780–990 is important for Sumatra's history, as it may have destroyed the camphor port known to travellers of the time as Fansur (thought to be modern Barus). However the most striking correlation between the two sites was for a layer immediately preceding the one which was created by the 2004 tsunami. This was dated to AD 1290–1400 in Aceh and AD 1300–1450 in South Thailand. Allowing for appropriate caution about the dating techniques and the possibilities of erosion and movements in the coastline, this suggests that the most definite tsunami of major scale comparable to 2004 in the Andaman-North Sumatra subduction zone occurred in the fourteenth century (Monecke et al. 2008; Bondevik 2008).

Another team centred at the California Institute of Technology and the new Earth Observatory Singapore has been analyzing the sudden uplift of coastal reefs in the chain of islands of the subduction zone, by dating the point at which their coral died by sudden elevation out of the water. By methods they call paleogeodesy or 'U-Th disequilibrium dating' of the death of coral, they provide dates much more specific, with a scope for error of only 3–18 years, than in the case of the carbon-14 dating of organic matter in the swales mentioned above. On the other hand the more sites they survey, the more complex the record becomes, with no exact correlation of major events between one site and another in the same general section of the subduction zone. Through these means the Caltech team showed in the Mentawai patch off Central Sumatra events they held to correlate with known earthquakes of 1797, 1833 and 2007, though with a surprising degree of certainty that the Mentawai earthquake of 2007 was far from extensive enough to release the pressure built up by subduction since 1797–1833. Further events, on a much larger scale than the Padang earthquake (only 7.6 M), can therefore be expected within the next decade or two (Sieh 2012).

Proceeding further back in time, these researchers have identified events of major uplifts of coral reefs in the Pagai Islands (southern end of the Mentawai chain off Padang) around the years (magnitude of uplift in brackets) 1347 (3.2 m), 1374 (>0.5 m), 1381 (1.3 m), 1607 (1.5 m), 1613 (1.2 m), 1668 (>0.5 m) and 1675 (0.8 m).[14] Supported by lesser events around similar dates, these dates support the idea that the west coast of central Sumatra would have suffered major earthquakes accompanied by traumatic tsunamis in the late fourteenth, the early seventeenth century, and the 1797–1833 period known to history, with quieter times in between these three phases.

The same team subsequently worked on Simeulue, the northernmost of the chain of offshore islands, the coral uplift of which reveals events that impact the west coast of Aceh as well as Nias and Simeulue itself. Their findings showed in one Simeulue site major uplifts in about 1394 and 1450 with a minor one about 1430, while another site further south showed major uplift (comparable to the 2005 Nias quake) in 1450, but little or none at the two earlier dates.[15] This matches very well the previously discussed evidence of swale sand deposits by the different teams, which identified the period 1290–1450 as the likeliest time for a major event on the scale of 2004. The Caltech results show that the big tsunami of northern Sumatra and the central Peninsula must have been in the latter half of that period, with probably a series of traumatic events between 1394 and 1450.

14 Sieh (2012). See also Kerry Sieh, Danny Natawidjaja et al. (2008:1674–1677).

15 Meltzner et al. (2010); Meltzner et al. (2009:409).

Revising Sumatran History

For the history of Sumatra this puts crucial new evidence on the table. We now know that at least one massive earthquake of around 9 M, and an attendant tsunami on the scale of that of 2004, must have accompanied the transition between Buddhist and Islamic Sumatra. This may help explain the seeming lack of continuity between the two eras, and our almost complete ignorance of the history of the earlier one. Almost nothing is understood about the societies which erected the megaliths of many highland locales, or the upland Buddhist temple complexes at (north to south) Padang Lawas, Muara Takus, Pagarruyung, Rambahan, and Karang Berahi.[16] The camphor-producing Batak area in the north and the gold-producing Minangkabau highlands of the centre when possible exported their products through the nearest ports on the west coast, which should therefore have been the entry points of Indian ideas such as Buddhism. Yet we have surprisingly little evidence of port-cities on the west coast before the Europeans, to rival the east coast river ports of Palembang (older Srivijaya), Muara Jambi, Inderagiri, Siak and Asahan in historic times, or the ceramic-laden site of Kota Cina near Medan in the twelfth-fourteenth centuries.

The only comparable ports on the west coast known to foreign travellers before historic times appear to have been Lamri (somewhere in the northwest corner of Sumatra), shadowy Daha or Barat (Calang), and Fansur (probably at Barus north of Sibolga). Even more mysterious are the clearly Indian place-names that remain on the west coast: Indrapura as a gateway for Kerinci in the extreme south of current West Sumatra province, and a triangle of half-remembered sites in the general area of Banda Aceh – Indrapuri in the interior, Indrapurwa on the coast to the west of the modern city, at a site now wholly submerged, and Indrapatra near the coast above Krung Raya to the east of the city. None of these, however, have left substantial physical remains of an Indian presence. The explanation must lie in part in the extremely volatile nature of the coastal region. Coastal settlements could not become established in the long term because they were disrupted every couple of centuries by severe earthquakes and tsunamis. Even when those port-cities did become established enough to build permanent monuments in stone, the evidence was ultimately obliterated by earthquake and tsunami. Lubuk Tua, the likely centre of the camphor port of Fansur, the most substantial west coast settlement known to Arab, Indian and European traders of the eight to thirteenth centuries, was carefully excavated in the 1990s, but the physical evidence of Indic religion

16 Current understanding of these sites is usefully summarized by John Miksic (2009).

there remains meager compared with the upland sites mentioned (Guillot 1998–2003). Minangkabau gold was an even older and richer Sumatran export than camphor, giving Sumatra its Sanskrit name *Suvarna-dvipa* (gold-land). If in the first millennium of the Common Era there was a port for its export somewhere on the west coast of central Sumatra, then the evidence for that has fallen victim to multiple seismic traumas in the interim.

The most substantial state of modern times, the sultanate of Aceh, arose rather quickly in the late fifteenth century and conquered the fertile northern coast to create a strong sultanate in the 1520s. Nurudd-din ar-Raniri begins the most reliable of the Aceh chronicles by saying simply that 'Sultan Ali Mughayat Syah [d.1530] was the first to rule the kingdom of Aceh', before which there were no rajas but only village communities (Nuru'd-din ar-Raniri 1966:31). In reality the trading centre of Lamri at the Northwestern tip of Sumatra was known to Arab traders to China from at least the ninth century, and to Marco Polo in 1292, while it sent missions to China in the period 1286–1412.[17] In the Aceh chronicle tradition, however, Lamri scarcely features as an antecedent.[18] Aceh appears a new beginning.

The knowledge that the Banda Aceh area is likely to have been hit by at least one major earthquake and tsunami in the period 1340–1450 helps greatly to explain this discontinuity. Can history help to make the moment more precise? Perhaps only to rule out the period between around 1400 and 1416, when the Ming reportage of the Zheng He visits to Sumatra makes no mention of a devastating tsunami. Ma Huan and other writers of this period make this to be an entirely Muslim place of about a thousand families, no pigs, copper coins, and dwellings 'the same as in the country of [Malayo-Muslim] Samudra'.[19]

This seems a quite different place from the Nan Wu-li described by Wang Dayuan in his 1349 account of travels over the previous two decades. His Lamri was clearly not Muslim or notably urban at all, and its people, naked above the waist, lived 'all over the hills'.[20] The new post-1400 Lamri must have been a settlement of Muslim traders from Samudra-Pasai and elsewhere. The pre-1349 one may have been a coastal outpost of a Buddhist capital further inland, or

17 G.E. Gerini (1909:698–705) provides a chronology of references to Lamri, generously adding some Chinese references to Ta-Shih. See also Tibbets (1979:138–140); Iskandar (1958:24–29).

18 There is a slight qualification to this general neglect. The 'Hikayat Aceh' written in celebration of Sultan Iskandar Muda (r.1607–1636), mentions among his ancestors one Munawar Shah, whose alleged exploits included being a descendent of Alexander the Great, spouse of a heavenly nymph, and 'king of Lamri' (Iskandar 1958:71).

19 Ma Huan (1970:122–124); Schlegel (1901:353–359).

20 Wang Dayuan, as translated in Rockhill (1915:148–149).

perhaps the remnant of a post-tsunami Lamri which had flourished earlier. A fourteenth century date for a tsunami appears therefore likely, making the Muslim Lamri of the Ming a relatively new entity taking advantage of China's intervention to claim the trading privileges that these missions brought. A post-1416 disaster, on the other hand, would help to explain why even Islamic Lamri was so little remembered in Aceh's origins,[21] and how a relatively minor injection of a few hundred Cham refugees from the Vietnamese conquest of their capital Vijaya in 1471 could establish such a distinctive imprint on the Acehnese language, which was spread along the northern coast by conquest in the 1520s (Reid 2006:7–8; Thurgood 1999:52, 47–58 passim).

The fragmentary historical record would therefore be compatible with evidence of two distinct mega-tsunamis affecting the Banda Aceh area, perhaps even at the two dates the coral uplift data from Simeulue has unexpectedly presented – 1394 and 1450.

The Uncertainty of Java

In Java, now the most heavily populated part of the Ring of Fire, the record is clearer for volcanic eruptions, both better documented in recent times and more likely to be noticed in the chronicles, than for earthquakes and tsunamis. Java's eighteen historically active strato-volcanos can be relied upon to produce spectacular eruptions every few years. The big eruptions of Explosivity Index (VEI) 4 and above, have occurred regularly in Java, at least at Galunggang (West Java) 1822 and 1982, Merapi (Central Java) 1872 and 2010, Kelut (East Java) 1826, 1919, 1951, 1966, 1990.[22] Each causes major local devastation and kills hundreds of people, though their broader effect on climate is harder to document with certainty. Apart from Krakatau (in the strait between West Java and South Sumatra) in 1883, rated VEI 6 for its destructive tsunami and the ash cloud that circled the earth, very little is known about how often the truly giant eruptions have occurred in Java.

As for the remoter past, firm data is scarce and speculation based on Javanese popular tradition tends to take its place. There is a tradition that Sumatra and Java were one island until a vast eruption separated them some 2,000 years ago. And although there is nothing specific to say so in the inscriptions, the disaster

21 The sudden disappearance of Lamri from Chinese and other records after 1430 was already noticed by Gerini (1909:682–683).

22 Smithsonian Institute Global Vulcanism Program, 'Large Holocene Eruptions', http://www.volcano.si.edu/world/largeeruptions.cfm, accessed 7 June 2012.

that overtook the first great Javanese civilization, the one that built Borobudur and Prambanan in the period 600–900 CE, is usually thought to have been a terrible eruption of Gunung Merapi. Certainly parts of the Mataram heartland (present Yogyakarta) were buried by an eruption at some time after the ninth century, including the Sambisari temple (8 km northeast of the centre of Modern Yogyakarta) recently discovered to have been buried under 6.5 metres of deposit. As discussed further in this volume by Jan Wisseman Christie, other disasters may have played a role in causing this fertile central Java region to fall silent for six centuries. A bigger eruption from a more distant volcano, less given to regular small eruptions than Merapi, may have caused the crops to fail for a year or more. Moreover a tsunami on the south coast cannot be ruled out as a major disruption of maritime access to Mataram.

An important survey of the historical evidence in 1987 pointed to the danger of the Sumatra subduction zone, but judged that by contrast Java appeared to be aseismic, with few serious earthquake or tsunami events on record (Newcomb and McCann 1987). This view, partly responsible for the inadequate research on Java's past, needs serious revision since the recent run of disasters, which includes two deadly tsunamis on the south coast of Java from deep earthquakes of only magnitude 7.8 (1994) and 7.7 (2006) respectively. For the record before 1900 most of the literature relies on Dutch reportage, although only after 1840, when a Dutch official was placed at the Southwest Java port of Cilacap, was there any Dutch presence at all on the endangered south coast. As we have seen elsewhere, pre-twentieth century Dutch sources seldom recorded natural disasters unless their own infrastructure, personnel or vital interests were affected. Even in the twentieth century the record is poorly documented, and before that significant tsunamis may have occurred without any record.

For earlier episodes Javanese sources must therefore be used, even though they are inadequately dated and generally opaque. They do make clear that the supernatural power of the southern ocean was even more important a theme of palace literature and ritual than that of the Merapi volcano, these being the twin poles, north and south, around which the cosmology of the central Javanese states have revolved. The chronicles of the Mataram kingdom of the seventeenth and eighteenth centuries, and the continuing ritual of its modern successors in Yogyakarta and Surakarta, make clear that both volcano and southern ocean are supernaturally dangerous unless propitiated and harnessed by rightful kings, but it is the Queen of the South Seas (Ratu Kidul) who is the more important figure. She is believed to have given the conquering founders of the Mataram dynasty, the shadowy Senopati (d. 1601) and mighty Sultan Agung (reigned 1613–46) their power to rule by her mystic union with them, though ordinary mortals who provoked her would be carried to a watery grave.

The most chronologically reliable of the early Javanese chronicles, the *Babad ing Sangkala*, does provide the most explicit evidence for a tsunami with words translated by Merle Ricklefs (1978:32–33) as 'In Mataram, they moved [the court] to Karta, indeed, when *disappearing, all was turned into sea*'. The italicized words, repeated three times in slightly different form for emphasis, are a chronogram (a date in words) for the Javanese year 1540, which began in February 1618 CE, at the beginning of Sultan Agung's campaigns to conquer Java. There is historic evidence, in other words, to locate a powerful earthquake and tsunami on the south coast of Java, which adds to the urgency of scientific research on the ground (Reid 2012).

Any large tsunami on Java's south coast would also affect Australia's northwest coast, as those of 1994 and 2006 did. Physical research in this sparsely-populated area is unfortunately hardly more advanced than in Indonesia, though one research team has cited geophysical evidence they argue indicates a truly mammoth tsunami in the Kimberleys at some time in the seventeenth century (Bryant et al. 2007).

***

What all of this suggests, is that in a zone as seismically active as Indonesia we must expect history to be discontinuous, through the effects both of volcanic eruptions alternately enabling and destroying intense agricultural production, and of tsunamis periodically destroying coastal settlements. This helps explain both the remarkably low population of the Indonesian-Philippines island arc as a whole before the nineteenth century, and the dearth of long-term settlements on the coastal plains of western Sumatra and southern Java.[23] Humans had flourished in this region for longer than in most of Asia, surviving the ice ages as was not possible further north, and enjoying environmental conditions congenial to human life and agriculture. The reason the populations of these archipelagoes remained much sparser than those in other parts of Asia before 1800 must be attributed in part to periodic disasters abruptly reversing the growth of agricultural populations.

23 The low population was previously explained in other terms, see Reid (1987:33–47).

CHAPTER 6

The *Longue Durée* in Filipino Demographic History

The Role of Fertility Prior to 1800

Linda Newson

Demographic trends reflect mortality, fertility and migration, which themselves are the outcome of complex social and biological processes. Examining the relative importance of these components of demographic change particularly over a *longue durée* is therefore a hazardous task. Even for Europe, where vital records are more readily available, there continues to be a debate about their relative importance (Wrigley 1969:146–202). In explaining long-term demographic trends in Southeast Asia, Boomgaard (2003a:198) has noted that scholars have tended to emphasise the role of mortality. High mortality has been invoked to explain low population levels in pre-European times and account for the initial decline in the native population following European contact, as well as the demographic crisis of the seventeenth century.[1] Furthermore, the subsequent phenomenal increase in population from the eighteenth century, though beginning in different regions at different times, has been attributed to falling mortality rates.[2]

Factors deemed to be significant in influencing levels of mortality in the Philippines, but also in Southeast Asia more generally, include changing levels of conflict, both inter-polity and with Europeans and Moros, the labour and production burdens imposed by this conflict and colonial rule, famines, epidemics and environmental hazards. Scholars that focus on mortality generally recognise that the death rate not only influenced demographic trends directly, but also indirectly through its impact on fertility, for example, through disrupting marital relations, undermining subsistence production and causing stress-induced amenorrhea. However, due to the shortage of evidence, especially for fertility, there has been little discussion about the relative importance of birth and death rates for historical periods. Writing in 1987, Norman Owen (1987a:56) could 'only lean toward the view' that Southeast Asian populations always had high fertility until the nineteenth century. He suggested that up until then, demographic growth had been tempered by periodic catastrophes, but thereafter reduced levels of conflict, the introduction of smallpox vaccination, famine

1 Reid (1987: 33–43, 1988, I:11–18, 1990b: 649–652, 2001: 46–49); Newson (2009: 60–61).

2 Owen (1987a:56); Reid (1987:43; 2001:50–53).

relief and improved sanitary conditions moderated their impact allowing the population to increase. However, focussing on the Philippines specifically, De Bevoise and Smith have argued that the death rate did not decline substantially in the nineteenth century. In fact, De Bevoise (1995:27) suggests that the impact of many acute infections was greater in the nineteenth century, when population growth and improved communications facilitated their spread (see also Newson 2009:262), while Peter Smith (1978:76) attributes raised mortality rates to the commercialization of the economy that resulted in deteriorating peasant livelihoods and increased susceptibility to disease. Owen (2002:10–11, 13) later recognised that he may have underestimated the importance of fertility, admitting that marriage may have occurred earlier, and that there may have been increased coital frequency and reduced birth control. Nevertheless, he still saw mortality as the most significant influence on population trends in the nineteenth century.

Any debate about the relative importance of mortality and fertility in determining demographic trends, especially for remote historical periods, is hindered by the lack of data and research, particularly on fertility. Whereas documentary sources often record the numbers dying in epidemics, conflict, famines, or environmental disasters, it is generally more difficult to find evidence, especially of a quantitative nature, for levels of fertility (Owen, 1987a:50). While by no means abundant, sources for the early Spanish Philippines are perhaps more available than for some other Southeast Asian regions since colonial rule was in the words of John Phelan (1959:13) 'essentially a missionary enterprise'. This meant that priests often resided in native communities for extended periods and moreover were interested in recording native customs and practices. Of particular importance are the seventeenth-century accounts of the Jesuit fathers, Pedro Chirino and Francisco Ignacio Alcina,[3] the latter serving in the Visayas for 35 years. Also, significant is the account by the Franciscan, Juan de Plasencia who was based among the Tagalog at Nagcarlan in Laguna de Bay and wrote in some detail on their law, customs and religion (Plasencia 1892).

It is not the aim of this chapter to argue that fertility was more important than mortality in determining demographic trends in historic Southeast Asia or even the Philippines alone, for it is unlikely that any generalization would be valid over such broad areas as these, but rather to suggest the relative lack of evidence should not lead to the dismissal of the role played by fertility. While quantitative data might be elusive, contemporary observations of birth

3 Chirino [1604] 1969. For the text of Francisco Ignacio Alcina's writings see: Martín-Meras, María Luisa and María Dolores Higueras 1974 and Yepes 1996.

and marriage practices do exist for the Philippines and some inferences can also be made from the changing nature of economic conditions and social practices recorded in colonial sources. It is important to note that these sources are only available for Luzon and the Visayas, and not for the southern islands of Mindanao and the Sulu Sea.

In as much as scholars have focussed on fertility there has been some debate as to whether changes in birth rates were related to economic or cultural factors. Much of the debate has centred on the nineteenth century for which records of vital events are more readily available. Reid has suggested for Southeast Asia as a whole that the adoption of world religions, especially Christianity and Islam, might have led to increases in fertility.[4] On the other hand, research on Indonesia, especially Java and Sulawesi, has suggested that fertility increases were associated with the commercialization of the economy through the increased demand for labour and/or opportunities to enhance household income, though views differ over the precise mechanisms underlying this relationship.[5] Did the birth rate increase because new employment opportunities encouraged women to have more children or did the heavy labour with which it was associated result in reduced breast-feeding, shortened intervals between births and hence larger families? It is not the purpose of this chapter to enter into this debate, but to suggest that the balance of cultural and economic factors influencing fertility is likely to have varied between different geographical regions and for different historical periods.

While Spanish colonial rule brought significant changes to life in the Philippines, the commercialization of the economy proceeded slowly. The islands lacked the gold and silver found in the Americas and there was little prospect of developing agricultural exports since their production was uneconomic given Spain's monopolistic control over trade and the long distance across the Pacific. Spanish possession of the islands depended on trade with China and commercial activities remained concentrated in Manila and its hinterland (Newson 2009:7–9, 36). Yet the Spanish had another objective in the Philippines, which was to 'civilize' its native population and convert it to Roman Catholicism. These processes brought significant changes to native economies and societies that were reflected in demographic trends.

Not only was the impact of Spanish colonial rule uneven in the Philippines, but it affected populations differentially according to the nature of the societies themselves and the environments in which they lived. My own research on Latin America has consistently suggested that marriage and birth practices

4 Reid (1987:40–41, 1988, I:16, 1993, II:161–164).

5 White (1973); Alexander (1984); Boomgaard (1989a); Henley (2005b, 2006, 2011).

played a key role in determining how quickly and to what extent native societies were able to recover from demographic crises occasioned by Old World epidemics and Spanish conquest (Newson 1995, 1998, 2006). For example, marriage rules governing the acceptability of particular spouses or remarriage that functioned to keep population levels low might threaten survival in conditions of demographic decline and reduced marriage pools (Newson 1998:55; Wagley 1951). As will be shown in this chapter, the imposition of colonial rule, including the introduction of Catholicism, directly affected cultural values, and as such requires consideration alongside economic explanations in understanding changing patterns of fertility, and hence demographic trends.

This chapter examines evidence for the role of fertility in demographic change in the pre-colonial and Spanish Philippines to about 1800. It examines differences in subsistence patterns and social practices that existed in the Philippines at the time Europeans arrived and argues that they are important in understanding long-term demographic trajectories in the islands under colonial rule, particularly as they affected the ability of societies to recover from the crises they experienced. In particular the chapter explores regional differences in age at marriage, bridewealth, penis inserts, abortion and infanticide, slavery and debt servitude that might have affected fertility. The focus of this chapter will be primarily on differences between societies in the Visayan islands and Luzon (the Tagalog region and Ilocos); it will not consider those islands in the southern Philippines that remained outside colonial rule or the semi-nomadic shifting cultivators in interior mountain regions for which evidence is scant before the nineteenth century.

Social Practices, Fertility, and Population Densities in the Pre-colonial Philippines

Social practices in the pre-colonial Philippines reflected the nature of societies found there. For the most part pre-colonial Filipinos resided in small communities, known as *barangays*, which were held together by kinship and common interest. The authority of their leaders, known as *datus*, depended on their ability to command popular support through demonstrated success in raiding, trading and regulating outside contacts, though it did not extend beyond their *barangays*. Relations between *barangays* were often hostile.[6] Characteristically settlements were located on the coast or a riverbank where the inhabitants

6 AGI, PAT 23–9, Miguel de Loarca [1582]; Chirino (1969: 363–364); Morga (1971: 271).

subsisted primarily on rice and fish.[7] A few larger settlements had emerged as a result of trading contacts with the Chinese and Arabs (Hutterer 1974:294–297; Junker 1999:22–23, 222–231), while Brunei traders had established towns at Pasig and Tondo in Manila Bay (Doeppers 1968:35–36). Nevertheless, even in the most densely settled regions of the Philippines, such the Central Plain of Luzon, around Manila Bay, and in Laguna de Bay, very few settlements exceeded a few thousand people. Population density in the Philippines was characteristically low, a feature that the islands shared with the rest of Southeast Asia. Yet, there were differences within the islands (Table 6.1). Population density in the Visayas was, with the exception of Panay, under 10 persons per square kilometre and in the Tagalog region about 15 persons per square kilometre, whereas in Ilocos population densities averaged about 27 persons per square kilometre (see also Xenos 1998:44). Clearly these figures are broad and indicative rather than precise.

Anthony Reid, in reviewing factors that might account for low population densities in Southeast Asia as a whole in the sixteenth century, suggests that while health and nutrition, age at marriage, family size, polygamy, abortion and infanticide might all play a role, these were less significant than

TABLE 6.1 *Estimated native populations in selected Philippine regions in 1565*

	Number of tributaries (*tributos*)	**Total population**	**Area in square kilometres**	**Persons per square kilometre**
Leyte	15,000	60,000	7,791	7.7
Samar	10,000	40,000	13,665	2.9
Bohol	6,250	25,000	3,913	6.4
Cebu	7,000	28,000	5,022	5.6
Panay	50,000	200,000	12,717	15.7
Negros	7,500	30,000	12,934	2.3
Total Visayas	95,750	383,000	56,042	6.8
Tagalog region (Laguna, Rizal, Cavite and Batangas)	30,000	120,000	8,241	14.6
Ilocos	43,000	172,000	6,465	26.6

SOURCE: NEWSON 2009:76, 140, 187, 256. AREAS ARE TAKEN FROM *CENSUS* 1905, II:28.

7 AGI, PAT 23-7, Diego de Artieda, no date; Doeppers (1968:29–30); Fox (1979:55–61); Morga (1971:270).

inter-polity warfare (Reid 1987:33–47, 1988:15–18). It should be noted, however, that Reid's argument does not focus exclusively on those killed in conflict, but does consider its impact on demands for labour, subsistence production, marital relations, and stress-induced infertility, which might have implications for fertility levels. That said, there were other influences on fertility that were less directly related to conflict. These included age at marriage, bridewealth, abortion and infanticide, penis inserts, and slavery and debt servitude.

Age at Marriage

Age at marriage is generally regarded as having the greatest cumulative effect on fertility in societies practicing little or no contraception (Marcy 1981:309). Unfortunately, little evidence is available for age at marriage for the Philippines and that which exists inevitably describes a post-contact situation where social practices might have been modified by attempts by both the secular and ecclesiastical authorities to encourage early marriage in order to expand the numbers of tribute payers and prevent what was perceived to be 'immoral' behaviour. Despite the fact that in colonial times the marriage of minors was expressly forbidden by law,[8] Father Alcina observed that in the Visayas many women married at twelve, though the marriage might not be consummated until a later date (Martín-Meras et al. 1974:205). Marriages among the Tagalogs of Luzon were also contracted at an early date, though what is described as marriage may have in fact been betrothal. Nevertheless, it is clear that sexual relations began at an early age. Writing in the 1590s Antonio de Morga observed that from their earliest years girls and boys had sexual relations and mixed freely with little modesty (Morga 1971:277). In fact they were described in the Boxer Codex as being more 'depraved in matters of the flesh and indulge in it more than Visayans' (Quirino and Garcia 1958:373, 427). Although the conclusion is drawn from European sources, these observations date from the 1590s and were therefore written only 25 to 30 years after contact. This limited evidence from the Philippines thus tends to support Reid's view that in Southeast Asia, 'marriage occurred early and almost universally' (Reid 1988:15, 158–160). It differs from Boomgaard's finding (2003a:198–203) for regions of Indonesia outside Java, where he argues that the existence of high bridewealth was conducive to late female marriage. Even if marriage was contracted at an early age, as will be elaborated below, the potential to raise large families, especially in the Visayas, was often not realized due abortion and infanticide.

8 *Recopilación* (1943), II: lib 6 tit. 1 ley 3: 190 real cédula, 17 April 1581.

Bridewealth

It is generally recognized that age at marriage, if not betrothal, can be influenced by the need to acquire bridewealth or, in other societies, dowry. Bridewealth was the norm in the Philippines, but there were variations in the level of bridewealth demanded. In the Visayas, marriages between people without property only involved the presentation of a few simple gifts,[9] but the marriage of a *datu* was accompanied by the payment of bridewealth, usually in the form slaves, gold or other valuable commodities (Chirino 1969:319–322; Yepes 1996:231–236). Slaves were particularly highly valued since land in the Visayas was inalienable so the key to socio-economic advancement was through the control of labour rather than land (Yepes 1996:207). For the marriage of the daughter of a *datu*, bridewealth might amount to the equivalent of 100 pesos. This was a substantial amount considering that a single slave cost between 6 and 12 pesos.[10] However, even within the Visayas there was a difference in the amount of bridewealth paid. Father Alcina noted that in Leyte, Samar and Ibabao bridewealth was given to the girl's father who distributed it among his *closest* relatives, whereas in other islands such as Bohol and Cebu it had to provide for *all* the relatives. Hence it was said that in the latter islands bridewealth was higher and that it acted as constraint on marriage (Yepes 1996:242). It might be suggested that this difference in the level of bridewealth might have contributed to the slightly lower population densities in the latter islands compared to the eastern Visayas (Table 6.1), though clearly this was not the only factor affecting population levels. There was no cultural or legal barrier to polygamy, but because of the financial burden that bridewealth imposed it was relatively uncommon and where practiced only involved two or three wives.[11]

Bridewealth was also payable in Tagalog societies, though in this case even fewer details are available for the pre-Spanish period (Morga 1971:274–275). The fact that polygamy was not practiced and that men had concubines suggests that bridewealth may also have been an obstacle to *formal* marriage there (Chirino 1969:81, 319; Morga 1971:277). Likewise, based on marriages practices today, it is likely that bridewealth was also paid in Ilocos in pre-Spanish times (Jocano 1982:170–172), but there is a lack of evidence both for the present day and for historical periods of the level required or of it being an obstacle to marriage. The overall impression from the accounts is that bridewealth may not have affected *age at marriage*, but whether people married at all. Prior to the

9 AGI, PAT 23–9, Miguel de Loarca [1582]; W.H. Scott (1994:140).

10 AGI, PAT 23–9, Miguel de Loarca [1582].

11 AGI, PAT 23–9, Miguel de Loarca [1582]; Pigafetta (1995:58); Yepes (1996:253).

mid-eighteenth century about 30% of the adult population in Jesuit towns of Tondo, excluding youths, was unmarried (Newson 2009:129).

Abortion and Infanticide

Related to age at marriage and bridewealth were abortion and infanticide, though the link may not be obvious.[12] Technically infanticide should be considered as affecting the death rate rather than the birth rate. Miguel de Loarca, a long-term resident of Panay recorded that in the Visayas:

> It is considered a disgrace among them to have many children, for they say that when the property is to be divided among all the children, they will all be poor, and that it is better to have one child. They are very strict in who they marry for nobody marries someone who is not of equivalent status.[13]

Similarly the Boxer Codex recounts how Visayan women wanted small families because they thought that 'in having many children they are like pigs', presumably because they would be poor, so that once a woman had borne two or three children she practiced abortion. Abortions were also common for unmarried women for whom it was a disgrace to become pregnant.[14] Infanticide was also practiced where families were too poor to raise more children, as well as for infants who were born blind or with deformities.[15]

Despite abortion and infanticide, Visayan beliefs reveal a concern for maternal and infant well-being. Pregnant women abstained from certain foods, such as turtles, tortoises, particular fish, venison or plantains, and after birth other taboos and practices were observed to strengthen the infant (Yepes 1996:110). Early colonial sources give the clear impression that abortion and infanticide were aimed at avoiding poverty and achieving social advancement (W.H. Scott 1994:118). This accords with Mary Douglas's (1966:267–268) observation that in traditional societies population control is often practiced to ensure access to status and prestige goods rather than as a response to shortages of basic resources. As such, the Spanish regarded the Visayans as selfish (Quirino and

12 For a brief review of historical sources for abortion and infanticide in the Philippines see Pedrosa (1983).

13 AGI, PAT 23–9, Miguel de Loarca [1582].

14 Quirino and García (1958:413); Rodríguez (1965–1988), XIV:481 Martín de Rada 16 July 1577.

15 Rodríguez (1965–1988), XIV:481 Martín de Rada 16 July 1577; Colin (1900–1902, II:149).

García 1958:413). Owen (1987b:11–12) has questioned the universality of abortion and infanticide in Southeast Asia as whole and, even where practiced, whether they were demographically significant. However, for the Visayas at least, Father Alcina judged that they were 'not the least of the causes of the decline and the lack of people' (Yepes 1996:143–144).

Abortion and infanticide appear to have been less common in Tagalog society. Although infanticide was routinely practiced for children born to unmarried women, since it was considered a disgrace (Quirino and Garcia 1958:373, 427), it does not appear to have been as widespread as in the Visayas. Tagalogs considered it desirable to have large families. This is evident from superstitions that encouraged fertility and the survival of infants. For example, a woman who wished to become pregnant raised pigs that were fed with the best foods and once the child was born they abstained from milling rice under the houses because it was thought that if the rice fell from the mortar and was subsequently consumed by a chicken the infant would die.[16] However, multiple births were considered undesirable and as such pregnant women abstained from eating any foods that were joined together, such as two bananas (Quirino and Garcia 1958:386, 442). Prayers were said for successful childbirth and it was thought that any woman or child who died in childbirth would be punished in the afterlife (Plasencia 1892:600, 603). Since children were highly valued, they were pampered, indulged and rarely punished (Quirino and Garcia 1958:386, 443).

In common with the Tagalogs, the Ilocanos, according to Miguel de Loarca, did not practice infanticide unlike their neighbours, the Pangasinans, who he indicates did so to avoid impoverishment.[17] However, Gaspar de San Agustín (1975:624) noted that it was practiced by Ilocanos in certain circumstances, for example, when an infant was born with an umbilical cord around its neck, since it was said it would be unlucky or die very soon.

Peter Boomgaard has discussed the link between high bridewealth and low fertility in explaining the low population densities in colonial Indonesia. The link he posits is through delayed marriage and hence reduced family size (Boomgaard 2003a:200–202). However, this study suggests that in fact the impact of high bridewealth in the Philippines at least was not so much to delay marriage, though there is a suggestion it may have had some effect, but rather to encourage abortion and infanticide to achieve smaller families so that bridewealth might be accumulated more readily to enable marriage at an early age. A similar low fertility regime has been found by Henley (2005a:466) in Sulawesi.

16 Quirino and Garcia (1958:377–378, 434). It is not clear precisely which native groups held these beliefs.

17 AGI, PAT 23–9, Miguel de Loarca [1582].

It might partially explain the apparent paradox of early marriage and low population densities in the Visayas; in Ilocos where these birth control mechanisms were largely absent, early visitors to the Ilocos coast encountered dense populations (Table 6.1).

Penis Inserts

In addition to abortion and infanticide, a distinctive feature of sexual relations in the Visayas was the practice of penis piercing.[18] This practice had the potential to influence both fertility and fecundity. Penis piercing was done with a pin or ring of metal, ivory or bone that often had wheels, spurs or other protrusions attached. It was said to provide women with greater stimulation and give them great pleasure so that men who did not possess them were scorned. This practice has been interpreted as reflecting the greater autonomy and equality of women in Southeast Asian societies, which was also evident in property rights, marriage practices and high divorce rates (Brown et al. 1988:11–12, 20). In order to accommodate the pierced penises it was said that women had their vaginas opened from the age of six, a practice that may well have affected fecundity. Father Alcina estimated that 19 out of 20 women suffered from pains in the abdomen and 'ahogamiento de la madre' (a nervous illness affecting reproductive functioning), which were said to be an impediment to procreation. Although Alcina attributed the affliction to frequent bathing, this cause seems unlikely; it could have derived from the use of penis inserts (Yepes 1996:144). Antonio de Morga (1971:278) reported that penis inserts resulted in women sustaining injuries and shedding a lot of blood, which together with infections that might follow could have induced infecundity, or even death. It was also said that men might die of virulent ulcers caused through using penis pins (Martín-Meras et al. 1974:13). The early seventeenth-century observer Francesco Carletti claimed that their use also reduced the frequency of sexual intercourse and was used as a form of birth control (Carletti 1965:83–84), though this appears contrary to observations above that women enjoyed it. Whether or not they were adopted as a form of birth control, they are likely to have had a negative impact on fertility. In Borneo today fertility among the Iban, who use penis pins, is lower than among the Rungus who do not (Appell 1968:205).

18 CDIU (1885–1932), III:461, Miguel de Legazpi, no date; AGI, PAT 23–9, Miguel de Loarca [1582]; Cavendish (1890:42); Martín-Meras et al. (1974:13); Morga (1971: 278); Pigafetta (1995:58); Quirino and García (1958:417–418).

Penis inserts were largely confined to the Visayas. In Luzon, the only exception were some groups in the highlands around Laguna de Bay, where they inserted small balls of tar the size of chickpeas under the skin of the penis (Quirino and Garcia 1958:387, 444).

Slavery and Debt-bondage

David Henley has collected evidence from Sulawesi and elsewhere to suggest that slave-owning societies and the slaves themselves had lower birth rates, since in both cases there was little incentive for women to have children.[19] This observation is pertinent to the Philippines since slavery and debt-bondage were widespread in some regions. It is worth noting, however, that 'slaves' consisted of individuals of varying levels of economic and social standing, ranging from those who lived independently and only worked for their masters for a few days a week, to those who were chattel slaves.[20]

Slavery and debt-bondage were widespread in the Visayas where the power of chiefs was dependent on the number of subjects and slaves he could attract or acquire through his charisma, military prowess, or wealth.[21] It has already been noted that bridewealth was often paid in slaves and, as argued by Boomgaard (2008b:92) for Indonesia, this may have encouraged indebtedness and debt bondage. An esteemed *datu* might possess 100 to 200 dependents and his wife slightly fewer, but even ordinary *datus* commonly had 50 to 60.[22] In Tagalog society also the authority of chiefs derived from their military prowess and economic status, which was evidenced by the number of 'slaves' they owned. Here some chiefs reputedly possessed 100, 200 or 300 slaves (Colin 1900–1902:75–77).

As for Ilocos, there is no mention of slavery or debt-bondage. The Ilocanos were regarded as a peaceable people and the absence of warfare suggests it was not being used to acquire slaves. It might be speculated that the high population densities found in Ilocos might have obviated the need to resort to control labour through enslavement. On the other hand, the prevalence of slavery has

19 Henley (2005a:370–373, 2005b:365–368, 2006:320–321); see also Boomgaard (2007b:132, 2008b:92).

20 Aguilar (1998:68–69); W.H. Scott (1983:142–145, 1994:133–135); for other regions of Southeast Asia see Boomgaard (2008b:85–91); Reid (1983).

21 W.H. Scott (1994:129); for contemporary accounts, see Morga (1971: 271–272); Quirino and García (1958:407); Yepes (1996: 207).

22 AGI, PAT 25–45, Breve sumario 1593; Yepes (1996:207).

often been correlated with economic insecurity (Henley 2005b:367) and Ilocos was a relatively infertile region that suffered from regular food shortages and famines (Newson 2009:184–185). The absence of slavery and debt servitude in the region is therefore surprising. An alternative explanation, might be that Ilocano society was less stratified than in the Visayan, so that chiefs lacked access to resources that could be used either to support debt dependents in times of economic insecurity or would require the service of 'slaves' to make them productive or enhance their status. Although there is no documentary evidence from the Philippines to support the link between slavery and low fertility, it is supported by evidence from other parts of Southeast Asia. Differences in its prevalence may therefore have contributed to differences in population densities at the time the Spanish arrived.

Review of the Evidence for Pre-colonial Times

From this review, marriage appears to have been contracted at an early age, which should have promoted larger families. However, abortion and infanticide were widely practiced. In the Visayas there was a clear preference for smaller families, perhaps encouraged by high bridewealth. In addition, the practice of penis piercing may have affected fecundity. In Tagalog society abortion, infanticide and penis piercing were uncommon, but slavery and debt-servitude were widespread, as they were in the Visayas (See Table 6.2). All these factors are likely to have had a negative impact on fertility. In contrast, all these social practices were uncommon among Ilocanos and it was here that the Spanish encountered dense populations. Differences in social practices may not only have contributed to the geographical differences in population densities observed by visitors to the Philippines, but may have influenced the way in which societies in different regions responded to the economic, social and political changes that came with colonial rule.

TABLE 6.2 *Summary of marriage and birth practices*

	Low age at marriage	Bridewealth	Penis inserts	Abortion/ infanticide	Slavery/debt servitude
Visayas	Yes	High	Yes	Yes	Yes
Tagalogs	Yes	High	Limited	Limited	Yes
Ilocanos	?	Probably low	No	No	No

Social Practices and Fertility in the Colonial Spanish Philippines

Changes to social practices introduced in the early Spanish period generally promoted fertility, though their full demographic impact was obscured during the first two centuries by the high mortality associated with Spanish conquest and the Hispano-Dutch War. Many of the changes that affected fertility came about as a direct result of efforts by the Catholic Church to make marriage and birth practices consistent with Christian beliefs. At the same time, the suppression of slavery and the commercialization of the economy, which changed land and labour relations, eroded traditional social practices, though their impact was uneven.

Age at Marriage

In colonial times officials and priests would have encouraged early marriage to increase the number of tribute payers and prevent what they perceived to be 'immoral' behaviour. During the early Spanish period there is scant evidence for age at marriage, apart from the observations noted above. At the beginning of the seventeenth century married men were liable for tribute payment at age sixteen and women from age twenty (Hidalgo Nuchera 1995:208), which suggests that at least some marriages were occurring at an early age. However, there are indications that by the early nineteenth century age at marriage had risen slightly. At that time, the average age at marriage for women in Nagcarlan in southern Luzon was about 22 years and many did not marry (Xenos and Ng 1998:204–205). It might be speculated that this rise in the age at marriage reflected the commercialization of the economy in the Tagalog region. Here labour drafts and reduced access to land may have disrupted sexual relations and delayed marriages, though it should be noted that Nagcarlan itself remained a relatively rural remote community.

Bridewealth

Changes to bridewealth payments may have had a more significant impact on trends in fertility. In pre-Spanish times bridewealth was often paid in the form of slaves, which enabled chiefs to control labour and build political support. In colonial times, however, the Spanish sought to suppress slavery and the political and economic power of *datus* became more dependent on land ownership and on the hereditary authority invested in *cabezas de barangay* and *gobernadorcillos* by the colonial state (Newson 2009:26). This process is likely to have reduced the role that bridewealth played in economic and political relations and also moderated the amount and nature of goods demanded. With slavery being abolished, bridewealth payments were more often paid in gold, bells, and china plates or increasingly in cash. In the Indonesian archipelago as well, the items required as bridewealth were often trade goods or where populations were dense, land.[23] Nevertheless, in the 1660s Father Alcina noted that even though Visayans then had greater freedom to pursue their own livelihoods and acquire wealth than they had in pre-Spanish times, in reality many were still too poor to accumulate the bridewealth necessary for marriage; indeed he suggested that *principales*, rather than acquire several wives, often sought concubines for whom bridewealth was unnecessary (Colin 1900–1902:74; Yepes 1996:242, 253–255).

As late as 1815, the Bishop of Cebu in the Visayas was claiming that bridewealth was an obstacle to marriage such that many people lived together for years without getting married (Bayle 1949:393). At that time bridewealth generally amounted to more than 10 pesos and could be as high as 40 pesos. Nevertheless, this was a significantly lower than that paid in the sixteenth century when, as noted above, the bridewealth for a chief might amount to 100 pesos. Even so, the amount required still represented a significant outlay considering that annual tribute was 1.25 pesos.[24] Any impact that a reduction in bridewealth may have had on nuptuality and therefore fertility would have been limited. Evidence for the eastern Visayas (Leyte, Samar, Bohol), which came under Jesuit control, suggests that between 1659 and 1755 the average percentage of adults who were married rose only marginally from 62 to 67% (Newson 2009:109).

Unfortunately, no evidence has been found of changes in the level and nature of bridewealth payments in the colonial period in other regions. However, one might speculate that changes would have been more significant in the Tagalog region than in the Visayas or Ilocos, since there colonial rule brought more radical changes to political structures and land and labour relations.

Abortion, Infanticide and Penis Inserts

Priests sought to suppress abortion, infanticide, and the use of penis inserts, as being inconsistent with Christian beliefs. Despite attempts at suppression

23 Boomgaard (2003a:202–203); Henley (2005b: 360, 2006: 313–314)

24 Hidalgo Nuchera (1995:166–169, 191); AGI, AF 296, Oficiales reales, 13 July 1709.

(Pedrosa 1983:33–36), it is clear that both abortion and infanticide continued to be widely practiced, and in the mid-seventeenth century were still deemed to be important in holding back population growth (Yepes 1996:144). Father Alcina claimed low fertility was due to unbalanced sex ratios, with most pueblos having twice as many women as men and many women remaining unmarried. This he attributed to men being drafted to serve in the navy, gun factories, and other public works (Yepes 1996:246). However, evidence, albeit to be used with caution, from the missions under Jesuit control in Leyte, Samar and Bohol reveals that this imbalance in the sex ratio extended to children. There were on average between 94 and 98 boys for every 100 girls, with the imbalance being greater in the seventeenth century when there were generally under 90 boys for every 100 girls (Newson 2009:92–93). Father Alcina noted that girls were more valued for the bridewealth they could attract, whereas boys were perceived to be a greater financial burden (Yepes 1996:233). These figures therefore hint at the possibility of differential infanticide for male infants, which perhaps became less prevalent in the late eighteenth century.

At the same time, Father Alcina reported that the use of penis inserts had largely died out (Martín-Meras et al. 1974:13). One might doubt the veracity of Alcina's account and suspect that the practice continued in secret. However, there is no evidence for their use after the mid-seventeenth century and they are not used in the Visayas today, although the practice persists in some other Southeast Asian islands, notably in Borneo. Since modern accounts suggest that the practice has a negative influence on fertility, it might be expected that its abandonment in the Visayas, for whatever reason, might have had a positive impact on fertility.

Slavery and Debt-servitude

The Spanish were opposed to slavery and sought to reduce levels of debt-dependency. By the 1660s it was said that *datus* who had previously owned 100 or more slaves by then had 10, and those who had possessed 50 or 60 had only two (Yepes 1996:207). Father Alcina's observation of the declining number of persons registered as slaves is generally supported by more detailed records of communities presided over by the Jesuits in the eastern Visayas, though slaves are not recorded consistently in these sources (Newson 2009:92–93). With their power over labour weakened, *datus* sought to retain their economic and political status by controlling it indirectly through the ownership of land (Aguilar 1998:71–75).

Unfortunately, there is little information on debt-slavery in the Tagalog region in colonial times. However, it may be speculated that the greater commercialization of the economy there might have undermined traditional socio-political structures based on slavery and debt dependence that had restricted fertility. For Sulawesi, Henley (2005b:368–369, 2006:320–323) has argued that the expansion of commercial activities provided economic opportunities and improved incomes, which enhanced economic security and obviated the need for labourers to resort to debt-bondage. However, in a later paper examining the link between the commercialization of the economy and raised fertility, he emphasizes the role of increased labour demands resulting in reduced breast-feeding and shortened intervals between births, rather than the decline in debt service (Henley 2011:7–8). For the Philippines, however, it could be argued that the increased commercialization in the Tagalog region brought increased impoverishment (Peter Smith 1978:76), which might actually have favoured the persistence of debt-relations and stimulated migration that delayed marriages and disrupted sexual relations. However, in the late nineteenth century, at Nagcarlan at least, age at marriage seems to have been falling (Xenos and Ng 1998:205).

The Longue Durée in the Philippine Demographic History: Accounting for Regional Differences

It is not possible to isolate fertility from other influences on demographic change, but this chapter speculates that regional differences in marriage and birth practices in the Philippines may have contributed to regional differences in pre-Spanish population levels as well as demographic trends during the colonial period.

Marriage and birth practices that controlled population growth seem to have been largely absent from Ilocos, where the Spanish found some of the highest population densities in the islands. Their absence may also partially explain the relative ease with which, after the dramatic decline in the early conquest period and occasional mortality crises, the population there was able to recover relatively quickly once economic and social stability had been achieved (De Bevoise 1995:22). In the second half of the eighteenth century the tributary population was growing at 1.8% a year (Table 6.3). While part of this population growth reflected more effective enumeration by Spanish officials, it cannot account for the scale of the increase. This growth put pressure on land and fuelled outmigration, especially by men, that ultimately in the twentieth

TABLE 6.3 *Growth in the Number of Tribute Payers according to Fiscal Accounts, 1591 to 1792*

Jurisdiction	Growth rate 1591–1655	Growth rate 1655–1700	Growth rate 1700–1750	Growth rate 1750–1792
Laguna (Tagalog)	−0.8	0.3	0.8	0.2
Ilocos	−0.8	−0.1	1.3	1.8
Leyte and Samar	−0.7	0.6	1.9	−0.2
Panay and Negros	−0.1	−1.3	1.6	1.0
Cebu	0.0	1.5	1.6	0.6
Total Visayas	*0.0*	*−0.8*	*1.7*	*0.6*

SOURCE: NEWSON 2009:311–312.

Note: These figures are based on the numbers of tribute payers, so they reflect in part the effectiveness of registration. They should be used as indicative rather than definitive.

century led to rising age at marriage and high female celibacy (Xenos 1998:46–48, 53–58; Newson 2009:200).

In contrast, in pre-Spanish times the Visayas were characterised by low population densities. Despite efforts by the missionary orders to suppress marriage and birth practices that were inconsistent with Christian beliefs, their efforts were protracted and only partially effective. These practices, which had functioned to control population growth, made it particularly difficult for societies there to rebound from economic, political and environmental crises. Also, this region did not experience a high level of commercialization that might have encouraged labour-demand led increases in fertility. As such, although the population recovered after the Hispano-Dutch war, population growth there was much more sluggish. With the exception of Cebu for which little evidence of birth and marriage practices exists, growth was especially low in the seventeenth century, where some islands registered declines. According to the registration of tribute payers, the population in the Visayas began to increase in the early eighteenth century. This was probably related in part to more effective registration, but may also have reflected the greater effectiveness of Jesuit teaching in changing marriage and birth practices. However, the growth was not sustained especially in Leyte and Samar, where in the second half of the eighteenth century the population appears to have declined. Part of the decrease would have been due to the high mortality caused by Moro raids, which affected those islands particularly badly, or reduced Jesuit supervision following their expulsion, which permitted resurgence in native practices.

However, it is worth noting that these islands also experienced fewer changes to agricultural practices and commercial expansion (Newson 2009:105, 111).

As for the Tagalog region, Manila's demand for labour and provisions fuelled the establishment of friar estates in its hinterland. This brought significant transformation to social and economic structures at the same that colonial rule undermined the basis of native authority based on the control of labour through slavery and debt-bondage. These processes together with the absence of strong birth control mechanisms should have encouraged higher fertility and favoured a rapid expansion in the population. However, the economic transformations that occurred in this region reduced access to land, created a demand for labour in Manila, and stimulated migration, which would have resulted in delayed marriages and disrupted sexual relations. Despite reduced controls on fertility therefore, population growth in this region was limited. These reflections are highly speculative and require further investigation, but they appear inconsistent with the arguments by Boomgaard and Henley that would link increased fertility to the demand for labour. However, if valid, they suggest that in arguing for a link between the commercialization of an economy and increased fertility, greater attention needs to be paid to the specific way in which commercialization affected land-labour relations and incomes.

From the last comments it will be evident that fertility cannot be considered in isolation, for demographic trends need to take account of mortality, migration and wider social and economic changes. However, from a practical perspective, analysing fertility separately can pin-point the precise mechanisms that regulate it and facilitate cross-cultural comparisons over time and space. In addition, this chapter suggests that a focus on regions where different cultural and environmental conditions prevailed can tease out some factors important in explaining demographic trends that might go undetected when they are viewed on a larger geographical scale. It also suggests that while there may be a demonstrable link between fertility and economic change, its significance should not be allowed to obscure other factors, such as broader cultural changes including those to religious beliefs, perhaps most evident in a colonial context, which may be important in understanding fertility trends especially over a *longue durée*.

CHAPTER 7

Glimpsing Southeast Asian *Naturalia* in Global Trade, c. 300 BCE–1600 CE

Raquel A.G. Reyes

The Loris and the Lory in Renaissance Europe

Sometime in the sixteenth century two remarkable creatures from Southeast Asia arrived in Europe. The slow loris, *Nycticebus coucang*, is a small, densely furred, tree-dwelling primate. It is nocturnal, foraging in the darkness for food using its large round eyes. Unusually for a mammal it is venomous, repelling predators with a poisonous bite and protecting its young by licking it with its tongue, thus coating it in toxic saliva. The purple-naped lory, *Lorius domicella*, is a species of parrot from the forests of the Moluccas. It is 11 inches long and a carnival of colour and sound: emerald green wings are set against deep red plumage, blue feathers cover its thighs; it has a curved orange beak, a crown of black shading to purple at the nape, hence its name, and, unlike lesser birds that utter pitched shrieks, the lory whistles a melodious 'weee ooo weee auuh' as it flits through the forest. We know something of the European adventures of the loris and the lory. The loris was presented to the imperial court of Rudolf II in Prague and drawn by the court painter to the Habsburgs, Giuseppe Arcimboldo (1526–1593), an artist and naturalist renowned for his painterly invention of composite heads. The purple-naped lory appeared in the painting *The Baptism of Christ* by the Venetian painter Giovanni Bellini (1430–1516), an arresting little detail in which the divine and the worldly came together for visual contemplation.

How Arcimboldo used his drawing of the slow loris is uncertain. The extant drawing shown here (Fig. 7.1. Top animal) is likely to have been based on a prototype that in turn was copied.[1] Whether he intended to incorporate it in a portrait or as a nature study is open to speculation. Arcimboldo was a court

1 The miniatures of animals and birds found in folio Cod. Min. 129, fol. 38r in the Österreichische Nationalbibliothek, in which the drawing of the slow loris is to be found, are copies attributed to Dietrich de Quade van Ravesteyn, who was in Imperial Service 1587–1609 and so roughly contemporary with Arcimboldo. I am grateful to Thomas DaCosta Kaufmann for this point and to Mag Prokop, archivist at the Österreichische Nationalbibliothek. See also DaCosta Kaufmann (2010:138).

FIGURE 7.1 *Slow loris (animal above).*
REPRODUCED WITH KIND PERMISSION FROM ÖSTERREICHISCHE NATIONALBIBLIOTHEK

painter but he was also a respected expert in the collection and depiction of *naturalia*. He was in direct contact with court physicians and naturalists throughout Europe, notably Franciscus de Paduanis and Ulisse Aldrovandi in Italy to whom he sent drawings of exotic flora and fauna. His nature paintings,

FIGURE 7.2 *Altarpiece, Church of Santa Corona, Vicenza, Italy.*
PHOTOGRAPH TAKEN BY THE AUTHOR

FIGURE 7.3 *Altarpiece detail.*
PHOTOGRAPH TAKEN BY THE AUTHOR

executed in gouache and watercolour, were probably drawn *ad vivir*, from life (Swan 1995:353–372). These works were not merely preparatory studies for his artworks. They played an important role in the dissemination of knowledge about natural history by complementing the prized imperial menageries of actual birds and animals, and serving as references for naturalists (DaCosta Kaufmann 2010:122–124). The range and extent of Arcimboldo's animal studies is only just being uncovered; among the many other exotic creatures he is known to have drawn was the purple-naped lory that also attracted Bellini (DaCosta Kaufmann 2010:132).

Bellini's painting is a startlingly exotic presence in a picture famous for its allegorical depiction of the history of Salvation (Fig. 7.2). According to Oskar Bätschmann, Bellini sought to illustrate the union of the Trinity and the Incarnation, and the earthly consequences of these events. The purple-naped lory is shown beside the standing Christ, positioned somewhat lower than knee level (Fig. 7.3); it is perched daintily upon a branch, a vivid flashpoint of red and green against a background of dark sandy rocks and shadowy crevices. Its presence in the picture can be variously interpreted. A parrot was thought to herald Caesar; or, in Christian symbolism, promulgated the arrival of God's Son or alluded to his earthly status. It is this latter meaning that most resounds. Captured from a distant land and brought to Europe via the Arabian import trade, the lory is directly tied to the world of trade and the collection of rare exotica. Placed in such close proximity to Christ, its presence is a reminder of the artist's intention to depict the event of Christ's baptism as a revelation of His dual nature.[2]

In a book that honours the pioneering work of Peter Boomgaard in the fields of economic, demographic, and environmental histories of Southeast Asia, this contributory chapter engages with one pre-eminent concern of that scholarship, namely the historical relationship between humans and their natural environment in maritime Southeast Asia in the *longue durée*. Over the last two decades, Peter Boomgaard's many publications on this theme have focused on the long histories of exploitation and processes of extraction of natural resources from the region's forests, seas, lakes, and rivers, and the effects these have had on the economies and environments of local societies.[3] Broadly and baldly stated, issues of depletion and sustainability have formed the underlying *raison d'être* in much of his work. The importance of this is undoubted – the region is today confronted with alarming levels of deforestation, the widespread destruction of marine ecosystems, and animals close to extinction. Southeast Asian rhinos, notably the two-horned *Dicerorhinus sumatrensis*, for instance, are today critically endangered and number fewer than 250.[4] Given that their horn was sought out since at least the Qin Dynasty, the survival of the Sumatran rhino is rather an astounding feat. In speculating on the certainty of environmental loss, from forest cover to animal species, Boomgaard asks whether we are witness to 'the long goodbye?', the addition of the question mark striking perhaps a timid note of optimism (Boomgaard 2005b: 211–235).

In this chapter, I will examine one aspect of the relationship between humans and their environment – the human desire to collect things from the natural world. Humans have long viewed the natural realm as a source from which medicinals, precious rarities and marvellous curiosities could be obtained. Such things were not only treated as valuable trade commodities, but also collected and treasured for their intrinsic beauty and fascination. Since at least the High Middle Ages, exotic animals, plants, medicinals and unusual objects gathered from the natural world interested European elites. Strange natural phenomena – from the wondrous to the freakish and the monstrous – had

2 Bätschmann (2008:175). In a similar vein, the presence of a sulphur-crested cockatoo, thought to be either from Australia or Timor, in the *Madonna della Vittoria,* painted by Andrea Mantegna in 1496, has recently given rise to fresh speculation over early trade networks that may have spanned Australia or eastern Indonesia, China and Europe in the late fifteenth century. http://www.theguardian.com/world/2014/mar/19/cockatoo-perched-in-renaissance-painting-forces-rethink-of-history, accessed 3 April 2014.

3 See in particular Boomgaard (1998a:275–295, 1998b:21–35, 1999:45–78, 2005a:97–120, 2007c).

4 For data on the Sumatran rhino, see http://www.iucnredlist.org/details/6553/0.

been bound to religious discourses, to political debate, to science and medicine, and to alien lands. Often interpreted as portents for good or for ill, they were cherished as *mirabilia* and collected for the innate cosmic power they were believed to possess.[5] Hoarded by church ecclesiastics and royal personages in treasuries or *Schatzkammern*, *naturalia* were prized for their novelty, rarity, and their putative medical properties. Natural objects that were incised and mounted, gilded or embellished – rhinoceros horns, coconuts and nautilus shells – joined the category of exquisitely crafted and equally prized *artificialia* (exceptional objects wrought by hand), which also included Chinese porcelain that had travelled the long cross-continental medieval trade routes connecting Asia with Europe and the Mediterranean. Alongside the trade in bezoar stones, the tusks of narwhals (believed to be unicorn horns), and ambergris from the near East, were rubies and camphor from Pegu, nutmeg from Banda, and cloves from the Moluccas. Over the centuries trade in Asian goods transformed European cities. With Vasco da Gama's discovery of a direct sea route to India in 1498, Venice was eclipsed by Lisbon that rose to become an important European entrepôt by the end of the sixteenth century, rivalling Seville, Antwerp, and Genoa (Gschwend 2004:16).

The sixteenth and seventeenth centuries were distinguished by voyages of discovery, European expansion, burgeoning global trade and new attitudes toward nature and the study of natural history. Gardens were created for botanicals brought from distant places via expanding networks of exchange; collections of *naturalia* proliferated and, with them, paintings and drawings were commissioned to celebrate the acquisition of exotica and to boast of the magnificence of royal cabinets of wonders. The enthusiasm of Renaissance royal collectors for rare and precious exotica was not merely to showcase riches. Possessing *naturalia* underscored as well as publicly exhibited a very personal expression of power and political pre-eminence. These incitements are nowhere made more apparent than the resplendent Habsburg collections amassed in Vienna, Prague, Madrid, and Lisbon. The *recâmara* of Catherine of Austria (1507–1578), Queen of Portugal, Habsburg princess, and the greatest collector in Renaissance Portugal, held a vast and precious collection that reflected her personal passion for India and the Orient and declared her ruler-ship and status (Gschwend 1994:15).

Royal *Kunstkammer* or *Wunderkammer*, botanical gardens, imperial aviaries and menageries of rare and marvellous creatures made manifest a ruler's discerning taste and symbolically displayed an empire's dominion. Collecting *naturalia* catalogued nature's munificence and in some instances, came to be

5 Daston and Park (1998); Platt (1999); Campbell (2004).

related to humanist theories and the quest for universal and encyclopaedic knowledge, subjecting nature to close observation and interrogation. In Renaissance Europe, students of natural history moved beyond discussions of nature by the ancients – Pliny the Elder, Aristotle, Dioscorides and Theophrastus – and created dedicated museums transforming the study of nature into a modern discipline.[6] The investigation of nature also found intimacy with artisanal practices of manipulating natural materials. Early modern artisanal recipes, for instance, called for quantities of vermilion, mercury, and lizards in the manufacture and purification of gold. Through such hoary sounding alchemical processes evolved modern empirical practices of natural science (Pamela Smith 2010). In all this, trade and commerce were never far from the scene. For some, commerce was at the core of collecting endeavours. Dutch engagement in trade and conquest in Asia and the Caribbean married new ways of knowing the world with turning a handsome profit from the trade in rare and exotic plants and medicinals, seeds, and animals. The study of natural history, medicine and science, even the philosophical concept of objectivity, owed their developments to the collecting of *naturalia* as a fundamental commercial activity and enterprise in the Dutch Republic.[7]

During the Renaissance, the study of natural history and the concomitant appetite for collecting exotic *naturalia* became integral to the production of art and culture, the emergence of new philosophies, and the development of science and medicine. In turn, collecting fostered cultural trends, even a mania fuelled by cross-cultural contact and processes of exchange.[8] In this history, collecting practices of other times, other places and other peoples lie at the far margins. What would a history of collecting look like if it were to be broadened out and reframed to take into account the more extensive geographical scope and longer time span of collecting? This chapter proposes to widen the aperture. Long before the European contact, I shall show, Chinese and other Asians were taking an active interest in rarities and curiosities from Southeast Asia's forests and seas. Quantitatively, these exotica never figured large in inter-regional or global trade, but the processes of their collection, transmission and utilization had diverse cultural consequences both at their points of origin and at their destinations.

6 Ogilvie (2006); Pamela Smith (1996). On the representation of nature, see for instance Enenkel and Smith (2007); Jorink (2010); Barrera-Osorio (2006). On the role of *historia* or civil history in contextualizing natural history, see in particular Pomata and Siraisi (2005); and on the little studied animal menageries in Spain, see Gómez-Centuríon Jimenez (2009).

7 Cook (2007). On commerce and the circulation of medicine, see Cook and Walker (2013).

8 Literature on this is extensive, but see in particular Jardine et al., (1996); and Smith and Findlen (2002).

Collecting Naturalia in Southeast Asian Historiography

Drug prospecting, botanizing, and collecting in the region's mountains and forests preoccupied European botanists, naturalists, missioner-priests, physicians and apothecaries throughout the early modern period. Some of the most eminent natural investigators of the age were active in the Dutch East Indies – Jacobus Bontius, Georgius Everhardus Rumphius, Hendrik Adriaan van Rheede, and Willem ten Rhijne.[9] In the Philippines, Catholic priests compiled herbals, and wrote huge and sprawling works known as *historía* and *relación* that included long, detailed sections on plants and animals; the Jesuit lay brother Georg Josef Camel assiduously collected flora and fauna, sending specimens to the Royal Society in London and to the Dutch physician Willem Ten Rhijne in Batavia, amongst many others (Reyes 1999). The indefatigable Spanish and French naturalists, Antonio Pineda y Ramírez and Luis Neé, botanized all over Luzon and Mindanao in 1792 as part of the Malaspina expedition, one of the most fruitful circum-global voyages to have been sponsored by the Spanish Crown (Reyes 2013). Items of *naturalia* and curiosities regularly made the journey across to Europe as part of trans-oceanic shipments of commodities. Their presence can still be traced in such VOC records as the comprehensive listings provided by the *boekhoudergeneral* and from archaeological recoveries of sea-wrecks such as the Manila Galleon *Nuestra Señora de la Concepción*, that traversed the Manila-Acapulco route and carried, amongst other things of local origin, white and yellow beeswax and styrax benzoin, quantities of which were found still intact in their earthenware jars (Mathers et al., 1990:19, 434). During these long voyages, botanical exchanges are now known to have occurred. Ethno-botanists have shown how coconuts from the Philippines taken aboard Manila galleons were responsible for the introduction and spread of modern coconut populations in Mexico and southwards to Peru (Harris 2012). The return route introduced many New World plants to Southeast Asia including vanilla that reached the Philippines possibly from Guatemala or El Salvador (Lubinsky et al., 2008).

It is not then an easy matter to explain why collecting *naturalia* has been somewhat neglected in the histories of pre- and early modern Southeast Asia. Part of the answer perhaps lies in the way in which the field has remained relatively isolated from themes in European art history, science and medicine.[10] There is also the tendency of economic historians of Southeast Asia to approach

9 See, for instance Cook (2001, 2005).

10 From somewhat different angles but nonetheless provocative for our current purpose, Victor Lieberman has proposed a rethinking of area studies divisions that seeks to identify trends that draw together 'Europe' and 'Asia'. See Lieberman (1999).

natural things from one analytical direction – as products and commodities that were exchanged for the manufactured articles sought by Southeast Asians, particularly Indian textiles, iron-ware, Chinese silks and porcelain. Of predominant concern has been the early modern Asian trade in spices – nutmeg and its derivative mace, cloves, cardamom, and long pepper from the Malay Peninsula and the Indonesian archipelago, and the region's interactions with Europeans, principally Spanish, Portuguese, and Dutch. A cursory glance at a summarized listing of luxury and market goods that constituted intra-Asian and inter-continental trade in the seventeenth century, however, provides an enlarged picture of the wide range and breadth of traffic in natural things. The principal 'export products' of mainland and island Southeast Asia in that period alone featured a variety of botanical, mineral, and animal items. The region's forests yielded benzoin, resins, camphor, honey, musk, elephant ivory and rhino horn, deer hides, lacquer, gharuwood (also known as eagle wood or aloeswood), sandalwood, and sappanwood. Precious gems, gold, silver, and other metals, were mined from the ground; and pearls, ray skins, and tortoise-shell obtained from the sea (Boomgaard 2007b:148–149). To these things could be added a trade in cowry shells, edible bird's nests, mother-of-pearl, *tripang* (a type of sea slug), red coral, and kingfishers' feathers.

Curiosity about unfamiliar peoples, objects and distant places, and a sense of acquisitiveness – the desire to possess remarkable goods – is a universal impulse. Although this is implicitly recognized in scholarship, it is only in recent years that the paradigmatic framework of European collecting has been questioned and new attention given to 'collectors outside Europe, [to] the circulation of goods across cultures and vast geographical spaces, [or to] the ways in which collecting and collections helped shape conceptions of other cultures in the first truly global era' (Bleichmar and Mancall 2011:4). As new research has now begun to highlight, the desire to collect European objects was felt across Asia. All manner of European astronomical devices, mechanical clocks, mirrors, double-barrelled guns and other firearms were of interest to Chinese, Japanese, and Indian elites.[11] Aristocratic Thais sought to acquire scientific and mathematical instruments; and paintings and other works of art in the European style were prized by Chinese and Muslim households in the Dutch colonial city of Batavia, in Java (Benson 2011; DaCosta Kaufmann and North 2014). Acquisition of these objects changed the world-view of Asians and how they viewed themselves. This chapter casts light on the value given to *naturalia* over two millennia. I will show what things were sought out, and what sorts of information or knowledge about Southeast Asia were produced as a consequence.

11 See for instance Chaiklin (2003); and Screech (2002).

In Search of the Strange and the Precious: Chinese Trade with the Southern Seas since the Third Century BCE

Chinese interest in natural things, deemed as luxuries and originating from foreign places, can be traced back to antiquity. Since at least the third century BCE aromatic clove from the Moluccas, brought by Nan Yüeh traders of the southern provinces of China (modern day Guangdong, Guangxi and Yunnan) and northern Vietnam (including Hanoi), were used by courtiers of the emperor, who demanded that all those in his presence should sweeten their breath (Wheatley 1959:45). China's belligerent southward expansion and conquest of the Yüeh lands after 221 BCE were chiefly economically driven. It was well known that the river deltas around Hanoi and Canton were agriculturally fertile. But the trade in rhinoceros horns, elephant tusks, kingfisher feathers, and pearls was lucrative enough to warrant the deployment of five armies constituting a total of 500,000 men (Wang Gungwu 1958:8). Subsequent conciliatory tribute missions to the Han emperor by the Yüeh proffered only the most luxurious and precious, which included quantities of rhino horn, one thousand kingfisher feathers and cones filled with honey-soaked 'cinnamon-insects' to be enjoyed as a gastronomic delicacy (Wang Gungwu 1958:13).

But it was the western regions – central and western Asia – to which China turned. The Roman Orient, known in Chinese sources by the term *Ta-ch'in*, an area encompassing Syria, Egypt and Asia Minor (Hirth 1885:vi) was thought to have the most precious, luxurious and rare treasures, a reputation that remained undiminished for many centuries. Chinese accounts from the third century CE spoke of the area as a source for tree resins, coral, pearls, amber, gemstones, as well as a variety of manufactured goods from textiles to ornamental coloured glassware. It was said that the region was defined by an 'abundance of precious things' and imperial envoys, eunuch servants to the Han emperors, were instructed to offer fine silks and gold in exchange for 'bright pearls, glass, rare stones, and strange things' (Hirth 1885:42). The *Sung shu*, written around 500 AD, detailed some of the marvels – 'gems made of rhinoceros' [horns]', 'king-fishers' stones' (thought to be chrysoprase), 'serpent pearls', asbestos the 'wondrous fire cloth', and other 'innumerable varieties of these curiosities' (Hirth 1885:46). Not only were things from the western regions esteemed for their preciousness but also, as Taoists were inclined to believe, for their magical properties (Wolters 1967:40). Pearls, for instance, would be customarily placed in the mouths of the dead and specimens from India and the Persian Gulf were thought to have mysterious or magical qualities and considered particularly special. They entered China via Yüeh ports and were known as 'luminous-moon pearls' or 'night-shining pearls' that could be obtained, it was said, from a 'pearl tree' (Donkin 1998:68).

Naturalia from Southeast Asia may have been obtained during the early exploring missions undertaken during the reign of Emperor Wu (141–87 BCE), which ventured to lands known in the records as *Huang-chih*, presumed to be in various places – Kancipuram (Wade 2009a), the coastal areas of the Bay of Bengal, and sites in Indochina or Sumatra (Wheatley 1959:19; Wang Gungwu 1958:20). Described in the *Ch'ien Han Shu*, missions to and from *Huang-chi* took place during the first five years of the Christian era. One return mission offered as tribute a live rhinoceros but whether the beast originated from South India or Indochina is debated (Wang Gungwu 1958:24). The missions may have also intended to trade in the northern Moluccas and the Sulu Archipelago, the latter being the string of islands situated in present day southern Philippines (Wheatley 1959:19).

Collecting and the Nanhai Trade

Wang Gungwu has perceptively written that the China Seas became more like the Mediterranean after the tenth century.[12] Until about the third century CE, the Nanhai trade, which centred on the South China Sea, had largely been confined to South China and Indochina. However, in the period of the Three Kingdoms, a push was made toward the Nanhai countries and the Southern Seas. The establishment of a new court demanded luxuries and treasures while additional impetus came from the southern state of Wu, a kingdom in South China, which had been shut out from the Central and Western Asian trade routes and forced to seek new markets. The task was treated with urgency. Sailors were sent out to search for pearls, tortoise-shells and corals in neighbouring countries; more significantly, an embassy was despatched to Funan, then a kingdom that straddled the Gulf of Siam, southern Vietnam and the Malay Peninsula, where trade to India was forced to pass through the Straits of Malacca. Commercial expansion into the so-called 'southern barbarian kingdoms' produced important results. For the first time references were made to Java-Sumatra, Siam, Burma, the islands of Borneo, Banka and Billiton, and the Malay Peninsula (Wang Gungwu 1958:39). A book entitled *Record of Strange Things of the South* described the seasonal monsoons of the Indian Ocean, the lands where trading was established, and the appearance of local boats with sails made of leaves from trees (Wheatley 1959:20). An eye for the odd was not uncommon in Chinese literatures of the period. Animals and events in the

12 Wang Gungwu (2008:7). For discussions related to the Tonkin Gulf and the Braudelian comparison see Cooke et al., (2011).

order of the fantastical, the supernatural, and the strange, had become a vital feature in popular writings of the early third century BCE that linked such phenomena to cosmographical concepts.[13] They also played a political role. Telling of 'anomalies' served to define and demarcate the known and the unknown lands that lay in the empire's borderlands and beyond, 'us' and 'them' distinctions that were, as Jorge Flores has put it, 'absolutely vital between…what is normal, and the unknown, which is always strange'.[14]

Trade in the luxury goods of the Nanhai was particularly enriching. Shih Ch'ung was the governor of the central and south-central provinces of Hubei and Hunan. He also controlled the trade that passed through Hanoi and Canton to the city of Luoyang in central China, and made a considerable fortune sending his own men out trading and levying heavy taxes on the merchants and envoys who used the routes. Stories about his extravagance abounded. It was reported that in his household were many beautiful women who held in their mouths exquisite perfumes whose sweet scent 'was wafted by every breeze' whenever they talked or laughed. He had fragrant gharuwoods ground into fine powder and scattered over an ivory bed upon which the most beloved of his maidens would be asked to step, rewarding those who left the dust undisturbed with a string of five hundred pearls while enforcing strict diets on those who left behind their footprints. These proclivities aside, Shi Ch'ung was an avid collector of *naturalia* and amassed a collection that impressed even imperial visitors. He reputedly owned an extraordinary assortment of pearls, perfumes and scented woods. The coral trees in his collection were known to be particularly fabulous and of a kind never before seen in China – specimens three to four feet tall with unusual stems and branches of vivid hues (Wang Gungwu 1958:35, ff. 20). Where this coral might have come from is difficult to ascertain. Red coral, *Corallium nobile*, had reached the Chinese from the Roman orient since the Han period but, by Sung times, a species of shallow water precious coral resembling *Corallium japonicum*, varying from the palest pink to deep red in colour, could have come from the Philippine archipelago, notably Mindanao (Wheatley 1959:80). Gharuwood was one of the most important aromatics imported by Southeast Asia and this ostentatious entrepreneur and collector would have sought out the best quality from Cambodia or the northern areas of the Malay Peninsula. By the Sung period, gharuwood was also coming from southeast Sumatra, Java and the eastern archipelago but the quality was poorer and, it was discerned, the 'smell caught the breath and its taste was bitter and pungent' (Wheatley 1959:70).

---

13 See, for instance, Strassberg (2002).

14 Campany (1996); Flores (2007:558).

The Nanhai trade in precious *naturalia*, driven largely by the demands of courts and elites, lasted for about five centuries to the end of the Qin Dynasty (221–206 BCE) (Wang Gungwu 1958:113). From 420 to 589 CE, attention had shifted to tribute and trade in religious and ritual paraphernalia or 'holy things'[15] as a result of China's increased devotion to Buddhism. Chinese converts, such as the fifth century pilgrim Shih Fa-Hsien, and Hsüan–Tsang and I-Ching in the seventh century, sought to study the Sanskrit texts and made the long pilgrimage to India. Their return journeys took them by way of the Southern Seas and the Nanhai, sometimes stopping for long periods of time in Srivijaya, as I-Ching did. Records of their voyages provided valuable information on routes and navigation to the kingdoms of Southeast Asia (Wheatley 1961:108). Intensified communication with the Buddhist kingdoms of Southeast Asia brought luxuries for use in temples and ritual – Buddhist relics, painted stupas, perfumes and fragrant woods for incense in exchange for silks and brocades (Wang Gungwu 1958:54). But precious things that delighted the eye or provoked curiosity remained highly desired – diamond rings, red parrots, bezoar stones and turmeric, some of which originated from the Indian Ocean and were re-imports, were sent as tribute articles to the Liu Sung emperor by two western Indonesian kingdoms in 430 and 449 (Wolters 1967:78).

Southeast Asian Medicinals: Early Chinese Interest

Political stability and economic prosperity in China during the brief period of the Sui dynasty (589-618CE) led to a more systematic search for the strange and the precious in lands to the south. Imperial demands for natural things drove up the prices for animal skins, hides, bones, tusks, horns, and feathers – articles that were destined for the courts and palaces to be used for luxurious furnishings and embellishments. Pearls were even incorporated into the structures of palaces, and pearl-encrusted beams, columns and pillars were the height of luxurious ostentation and display (Donkin 1998:217). Aromatic tree resins from Southeast Asia were particularly prized as *materia medica*. Wolters has noted that camphor, known as '*P'o-lu* perfume' from northern Sumatra first made its appearance in Chinese records at the beginning of the sixth century when physicians recommended it to aid women in difficult childbirth: 'let them take a little camphor and grind it into a fine powder and swallow it down with newly drawn powder' (Wolters 1967:122). By the seventh

15 The phrase is Wang Gungwu's in Wang Gungwu (1958).

century benzoin and 'unicorn desiccate', or 'dragon's blood', the scarlet coloured resin obtained from the *Daemonorops* species found in western Indonesia, had reached China and were regarded as prestigious medicines, valued for their cleansing and fumigatory qualities. Unicorn desiccate was thought to work on 'the five viscera', and 'on evil air inside the body'. It was said to stop pain and break up 'accumulations of blood', and to heal ulcers by creating flesh (Wolters 1967:123).

Collecting Information during the T'ang

Despite growing demand and an increased range of Indonesian natural things reaching the Chinese market, it was not until the later T'ang period (618–907 CE) that the Chinese began methodically collecting information about the lands with which they had commercial dealings. Maps of the most important trade routes were drawn up and works such as a book titled *On the Esoteric Doctrine remitted from the South Seas* sketched out the chief routes to Southeast Asia (Wheatley 1959:21). T'ang histories containing descriptions of Malayan kingdoms written by Buddhist pilgrims en route to India, I-Ching being the most well-known, contributed valuable geographic and navigational knowledge of the South Seas and were also replete with descriptions of the local products and peoples, particularly the striking appearance of rulers. In the kingdom of *Tan-tan* (possibly Natuna Islands off the north-west coast of Borneo) for instance, we are told of a king whose body was perfumed with fragrant powder and whose country was abundant in sapanwood, betel-nut, hornbill, and peacock (Wheatley 1961:52). By the thirteenth century, trade between China and the Malay Peninsula concentrated on states that were rich sources of jungle products. The kingdom of *Tan-ma-ling*, thought to be a dependency of Srivijaya and a state in the Ligor district, offered beeswax, lakawood, gharuwood, ebony, camphor, ivory and rhinoceros horn for which foreign merchants exchanged pongee parasols and umbrellas, silk thread, rice, salt, sugar, porcelain vessels and earthenware bowls (Wheatley 1961:67); and *Teng-liu-mei*, possibly far to the north of the Peninsula, was known to have had excellent gharuwood and lac (Wheatley 1961:74).

In 623 and 624 CE, tribute missions were sent to China from Siam, the Malay Peninsula, Java, and various parts of Sumatra including Palembang and Jambi (Wang Gungwu 1958:74). So considerable was the T'ang trade in Southeast Asian aromatics and spices that their use for culinary preparation and genteel purposes – scenting clothing and bath water – had become common (Wheatley 1959:32). Jan Wisseman Christie (1998) suggests that with the expansion of Chinese trade in products from western maritime Southeast Asia and increased Southeast Asian interest in South Asian exports, the period from seventh to eight centuries marked the beginning of trade booms and depressions, a pattern of peaks and troughs that would last until the early fifteenth century.

Arabic Trade and Knowledge

Indian Ocean trade and commerce was significantly heightened by Arab merchants who sought out exotica from India, China, and the islands of the Malay archipelago, bartering Arabian gum resins, thorough-bred horses, cottons and other textiles, and metal objects for silk, sandalwood, porcelain, and black pepper (Islam n.d.; Tibbetts 2004). The increased presence of Arab merchants in Eastern markets is recorded in a diverse range of Arabic works, from travellers' accounts and geographical treatises to historical, medical and navigational works. Although some were derivative or plagiarized, the accounts amount to an Arabic conceptualization of Southeast Asia in an early period. The emphasis given to Southeast Asia's spices and drugs by Arab geographers reflected the interest in Southeast Asian products useful to Arabic medicine, but Arab writers also recounted the experiences of their voyages as well the routes that were taken, the local customs and the riches of the region. For the period between the start of the China trade and 851 AD, the most important were the travellers' accounts and navigational descriptions found in *Akhbār al-Şin wa'l-Hind* by Sulaymān Tājir and work by Abū Zaid that described places on the China route and trade expeditions to the Malay archipelago and the west coast of Sumatra. The popular collection of travellers' tales in *'Ajā'ib al-Hind*, dating from the end of the tenth century, were filled with fanciful stories of strange beasts and events of heroism, but also gave pragmatic details on trade voyages to the Malay Peninsula and west coast of Sumatra, as well as the activities of the Persian Gulf traders in the area (Tibbetts 1979:5–9). Few travellers passed up on the opportunity to collect exotic animals. One discerning Arab geographer and traveller, who made it a point to seek out creatures that possessed truly extraordinary qualities, was particularly impressed by the fabulously coloured feathered parrots found on 'Zabaj' or Aceh that could be taught, allegedly, to speak Arabic, Persian, Greek and Sanskrit.[16]

16 Abī Bakr Ahmad ibn Muhammad al-Hamādānī Ibn al-Faqīh (tenth century) writing in *Mukhtasar Kitāb al-Buldān* cited in Islam n.d.:17.

Naturalia of the Nam-Viet

Chinese information about Nam-Viet became more readily accessible in the latter half of the eighth century with the publication of numerous standard maps and gazetteers that Sung encyclopaedists came to depend upon. The landscapes, climate, natural history and ethnology of the region succeeded in capturing the imagination of administrators who, in the course of fulfilling a temporary posting in the south, produced more detailed observations and guide-books. Meng Kuan (writing in the ninth century) and Fang Ch'ien-li (who had an eye for the fantastic and liked to write love poetry to a pining concubine), were bureaucrats with a romantic bent. Their respective works *Study of the Strange Creatures of the Southern Quarter* and *Study of the Strange Creatures South of the Passes* focused on the natural world and recorded encounters with the fabulous inhabitants of the mountains, islands, and rivers (Schafer 1967:149). Demand for *naturalia* of the southern seas also resonated in T'ang literature, in which the exotic beauty, utility and alien charm of plants, animals and minerals of the southern seas came to be a distinctive feature. The shimmering plumage of kingfishers was described as 'halcyon' in romantic verse; feathers were woven into fabulous dresses to be worn by court beauties and placed on the same par as precious jewels. Bird collectors filled their gardens with specimens that were especially sought out not only for their chromatic splendour, but also for their purported ability to learn Chinese.

Rhinoceros did not have quite the same literary cachet, but the demand for their horn, as with pangolin scales, for medicinal purposes also remained high. Rhino horn proved a versatile remedy in Ming-era Chinese *materia medica.* It could act as

> an antidote to poisons…to cure devil possession…to remove hallucinations and bewitching nightmares…for typhoid, headache, and feverish colds. For carbuncles and boils full of pus. For intermittent fevers with delirium. To expel fear and anxiety, to calm the liver and clear the vision… An antidote to the evil miasma of hill streams. For infantile convulsions and dysentery.
>
> READ 1931.

Minerals and stones of fantastical shapes and hues had a place in both pharmacopeia and literature. Red sulphide mercury or the so-called 'cinnabar of Viet' was described in official T'ang *materia medica* as an authentic panacea for all manner of serious diseases (Read 1931:157). Stones were imagined to be transformed animals or sacred relics and revered in poetry and collected by avant-garde connoisseurs. The young poet Chang Hu indulged his melancholic and solitary disposition by composing verses in a garden he constructed from his collection of strange stones and 'stone bamboo shoots' from Nam-Viet (Schafer 1967:155).

Edible plants of the tropics also caught the imagination and were appreciated for their beauty and symbolic worth. Coconut shells were polished, painted and fashioned into cups and pitchers and admired as objects 'precious and rare, worthy of admiration' (Schafer 1967:174). Southeast Asian palms were propagated in northern climes and *pinang*, the Malay word for the areca palm, crossed over into Chinese usage as *pinlang*, an adaptation indicative of the extent to which the Southeast Asian custom of giving and chewing betel was accorded attention, though the fashion for chewing betel in northern cities seems to have stemmed from a belief in its medicinal value. Banana leaves, used in the manufacture of cordage and cloth, also possessed sensory appeal. T'ang scholars delighted in the plant's inflorescence, the broad, green, frayed-edged leaves and the soft rustling sounds they made. Green gardens of banana were cultivated solely for aesthetic enjoyment. For Liu Ch'ang, a ruler with a penchant for Persian girls, banana forest gardens were the preferred setting to while away the time with a favourite concubine (Schafer 1967:187).

Sung Commerce

Much scholarly attention has been given to interactions between Ming China and Southeast Asia, particularly in relation to the formidable naval expeditions undertaken by the legendary Admiral Zheng He from 1405 to 1433, and to events in Malacca and Indonesia.[17] However, it is to be recalled that the vitality of Chinese maritime trade with Southeast Asia and the Middle East reached unprecedented levels several centuries earlier, during the Sung period (960–1179 CE). Much is known of the sheer range and variety, value and volume of spices, aromatics and drugs, referred to in contemporary Chinese records as *hsiang yao*, that flowed into China from Tonkin, the Malay Peninsula, Siam, Java, Western Sumatra, Western Borneo and parts of the Philippine archipelago, via Arab and Hindu merchant networks in the Sung era.[18] Clove imports

17 See, for instance, Salmon and Ptak (2005); Suryadinata (2005); Wade and Laichen (2010).

18 For details on the pattern of long distance trade and the routes taken during the Sung and later Yuan and early Ming periods with regard to China's trade in cloves, see Ptak (1992, 1993).

were of only slight importance in the Sung period and remained so through the Yuan and early Ming dynasties (Ptak 1999:149). Late tenth-century Sung dynasty annals list rhino horns, ivory, carnelians, pearls, turtle and tortoise shells, incense and scented woods amongst the principal articles of trade. The quantities were not insignificant – inventories and lists compiled in the tenth and twelfth centuries show the number of imports running into several hundred items. Between 1049 and 1059 China's annual imports of elephant tusks, rhinoceros horns, pearls, aromatics reached 53,000 units; by 1175 the count jumped to over 500,000 units (Wheatley 1961:60).

From the tenth to the thirteenth century, the Malay region's pre-eminent entrepôt was the thalassocracy of Srivijaya. At Srivijaya, goods from other parts of Southeast Asia, India and the Middle East were transhipped for the Chinese market and, in return, Indian Ocean states were supplied with Chinese and Southeast Asian goods. Srivijaya was recognized as China's major trading partner and the number of tribute missions despatched – fourteen over a thirty-year period beginning in 960 – attests to the aggressiveness with which that status had been pursued. Gifts of aromatics, ivory, rosewater, and rhinoceros horns, coral and other high-value items from the region, the Middle East, and India, was a calculated demonstration of Srivijaya's access to and relations with major Asian trade zones and its sensitivity to the needs of the Chinese market (Heng 2009:78–79). In return, Srivijaya's elite envoys were presented with gifts of white stoneware, lacquer ware, silks, gold and silverwares, copper cash, headdresses and harnesses that acknowledged the missions had preferable status (Heng 2009:79). Ultimately, Srivijaya's success depended upon the maintenance of good relations with regional trading partners. The faltering supply of cloves, black pepper, and sandalwood incense from Java and the Moluccas in the late tenth century came as a consequence of a conflict between Srivijaya and Java as the latter jostled to supplant the former's trade monopoly by throttling the flow of products within its sphere of control.[19]

Chu-fan-Chi: 'Foreign Geographies' in China's Age of Renaissance

Chinese demand for exotic aromatics and drugs fostered efforts to understand their properties through observation and description. Commercial knowledge was invaluable, and Chinese traders were quick to develop multiple and complex classifications for aromatics that drew subtle distinctions between freshness,

19 Heng (2009:80). See also Meilink-Roelofsz (1962).

coloration, and smell, associating these qualities with place of origin, as in the case of gharuwood, camphor, and gum benjamin. But there also appeared erudite compendiums and treatises – Yeh T'ing-kuei's work on aromatics written in 1151 being a notable example – which described the appearance, sensory and medicinal properties, local uses, and vernacular names of objects and substances. A third source of information was what came to be known as 'foreign geography', which built on and contributed to commercial knowledge, in addition to providing a wealth of descriptive detail on foreign places, customs, and the natural things that had entered maritime trade. One of the most outstanding examples in this genre was the influential ten-volume work *Ling-wai Tai-ta,* compiled in 1178 by Chou Ch'u-fei, an assistant sub prefect, whose meticulous observations on the geography and commerce of Southeast Asia, India and the Middle East were plagiarized in subsequent works. The *Chu-fan-chi* by Chao Ju Kua, which appeared less than 50 years later, for instance, had been 'leavened with a sprinkling' of borrowings from its predecessor (Wheatley 1959:7–8). Such works, although poorly regarded by Confucian scholars as an obscure field, was part of a fecund era of intellectual, philosophical, and artistic endeavour that has been called China's 'Age of Renaissance' (Hirth 1896: 61).

The most valuable source for Sung traders was the *Chu-fan-chi* or 'Record of Foreign Countries' written by Chao Ju Kua and published as a handbook in 1226. But the work was more than a standard purveyor of information on foreign trade in Sung times. The first section is suffused with lively descriptions and observations of people and cultures of lands encountered by traders who operated along the sea-routes of Asia, coastal Africa and the Mediterranean. Indeed, the author seemed to delight in imparting details that would strike his reader as being unusual or provocative, even entertaining. It was reported, for instance, that the devout Buddhist king of Malabar employed a fellow to supervise and regulate everything he ate and, should the royal personage fall sick, the officer would be obliged to taste his faeces and 'treat him according as he finds them sweet or bitter' (Chao Ju Kua 1911:88). Though possessed of intimate information, Chao Ju Kua was no eyewitness. Rather, as Inspector or Commissioner of Foreign Trade in Fukien, Chao Ju Kua enjoyed privileged access to information on countries in South and East Asia through his personal association with foreign merchants and sailors, principally Arab, Persian and Indian traders who passed through Fukien.[20] He attended to reports of how, at Srivijaya, merchants from the Middle East gathered to barter with luxuries that

20 For a detailed assessment of the inspectorate office Chao Ju Kua occupied and his family lineage see Hirth (1896).

included rose-water, ambergris, aloes, myrrh, coral, cat's-eyes and camel's hair cloth (Chao Ju Kua 1911:116). He was alert to a host of curious *naturalia* and registered his interest in the strange. Bottled gardenia flowers brought by camel caravans exuded a scent that was appreciated by the Chinese for its exoticism (Chao Ju Kua 1911:134). He marvelled at reports of civet, red or brown coloured specimens possessed of long dog-like legs that 'moved as if it were flying' (Chao Ju Kua 1911:234); and made a note of *mo-so* stones or bezoar stones from Nan-p'i, a region that ostensibly extended from Cambay to Ceylon which served as an efficacious antidote for poison from reptiles. 'Worn in a finger ring', Chao Ju-Kua observed, 'if one is poisoned and licks it, one is at once cured, so it may well be considered a life preserver'. (Chao Ju Kua 1911:90.)

The second section of the *Chu-fan-chi* was a methodical account of the principal items to enter maritime trade with the most original descriptions relating to Southeast Asia. Spices, aromatics, and valuable woods are treated as staples from the Moluccas, the Lesser Sunda Islands, and Java, but the Indonesian archipelago and the Malay Peninsula also provided a range of rarities for which there was a ready market. As Chao Ju Kua recounted, rhinoceros horn was obtained from the northern part of the Malay Peninsula and Java, ostensibly from creatures with fierce and violent tempers that 'ran so fast one might imagine them to be flying' (Chao Ju Kua 1911:233). High quality beeswax, areca palm, abaca and pandanus fibres were supplied by Srivijaya, Borneo and the Philippines, and from central Java came unusual or strange edible fruits – bananas that grew monstrously long, and jack fruit, which Chao Ju Kua treated as a hybrid confusion of pumpkin, chestnut, and mandarin orange (Chao Ju Kua 1911:83). Interest in medicinals figured prominently and Chao Ju Kua was careful to note the places of origin and the healthful benefits of a host of natural substances that were known to locals. Styrax benzoin from Siam, Malacca, Sumatra, and Java was valued for its powerful scent (Chao Ju Kua 1911:202); sun-dried lychee from central Java was a cure for bowel complaints (Chao Ju Kua 1911:83); camphor, particularly Barus camphor obtained from Sumatra, Ligor and Patani, was esteemed for its superior quality, pungent aroma and medicinal properties by Malays who applied the substance on aching teeth, running or inflamed eyes and open sores (Chao Ju Kua 1911:194). Ginger from Cambodia was valued by the Chinese as a luxurious addition to the table, and the peel was sought after by Chinese apothecaries who used it for medicinal purposes (Chao Ju Kua 1911:53). The fauna of Southeast Asia did not escape Chao Ju Kua's notice and he reported how in Borneo, the Philippines, Java and islands of the eastern archipelago, tortoise-shell was obtained from tortoises 'caught on moonlight nights during the autumn' (Chao Ju Kua 1911:238). Birds with rainbow plumage or in dazzling white, of exceptional intelligence, were treated as novelties in China and Chao Ju Kua recalled choice specimens of parrots from Tonkin and lories from the southern Moluccas, some of which, so it was said, complained of being cold to the Emperor, who ordered them to be sent home (Chao Ju Kua 1911:236).

Southeast Asia's Ages of Commerce

Thus far we have seen some of the sorts of natural things from Southeast Asia that, for centuries, were sought out and valued by Chinese and Arab merchants and collectors as exotic rarities or luxuries, or appreciated for their intrinsic qualities, be they strange, novel, or medicinal. Southeast Asians, in return, developed an insatiable appetite for the things foreigners brought, which were largely manufactured objects. Chao Ju Kua related how Chinese traders attempted to satisfy local demands for ornamental glass and porcelain throughout the region. In Borneo and the Philippines, for instance, coloured glass beads and bottles were popular. While exports of trade porcelain were in demand from Annam to Malabar, specific areas demanded specific types. Blue and white porcelain were traded in Java, white in the eastern archipelago, and celadon in Borneo. Aside from porcelain, a brisk trade for cinnabar could be had in Java, where the women used the substance for dyeing silk clothing and as a cosmetic, specifically nail polish 'for dyeing the fingernails'. The demand for cinnabar was evidently high enough for it have been considered a trade staple (Chao Ju Kua 1911:83).

Archaeological evidence has shown that long preceding Chao Ju Kua's medieval reports, foreign objects had already begun to exert a profound impact on Southeast Asian lives, connecting their cultures, economies, and social structures with the wider world. Throughout mainland Southeast Asia, archaeological excavations of prehistoric villages along river valleys have unearthed copper and tin artefacts, and marine shell ornaments. Roman coins and seals, evidence of glass-making, iron-smelting and forging techniques that originated from India, all indicating the far-reaching extent of trade networks, the movement of ideas as well as goods, and the changing nature of contact, connection, and exchange that occurred since the fourth century BCE.[21]

As David Henley and Kwee Hui Kian in this volume (see Chapters 8 and 10 respectively) also discuss, historians understand Southeast Asia's 'ages of

21 See Glover and Bellwood (2004); Bellina et al., (2010); Manguin et al., (2011); Kulke et al., (2009).

commerce' as periods of major social and economic transformation.[22] These periods witnessed a host of revolutionary developments – from the emergence and cosmopolitanization of port cities as hubs of commerce to new forms of consumption and increased monetization. Trade formed the backbone to these changes. Commerce with Srivijaya introduced local societies to a diverse range of material objects, demand for which demonstrated growing cosmopolitan tastes and an appreciation for objects both natural and made by human hands. A tenth-century Indonesian-made vessel (known today as the Intan shipwreck) is understood to have been trading between Srivijaya and the Javanese state of Mataram, carrying Chinese and Indonesian bronze-ware including highly embellished mirrors and sculpture for religious and domestic use; thousands of pieces of ceramic ware manufactured at Yue, Guangdong and Fujian kilns; fine paste ware from southern Thailand; glass-ware and beads in an array of hues from the Middle East; tin from the Malay Peninsula; and gold jewellery wrought in Sumatra and set with rock crystals, amethysts, rubies and sapphires. Interest in organic material was not inconsiderable, and amongst the recovered artefacts were tiger teeth and bones, ivory, deer antlers, aromatic resins, and candlenuts (*Aleurites moluccana*), the latter probably intended for culinary and medicinal use.[23]

Throughout the late first millennium, the Philippine archipelago was linked to an international trading system that involved Borneo, Java, the Moluccas, Champa, and China. It supplied gold ore, forest and marine products in exchange for ceramics and silk, principally Chinese, and base and precious metals. From the tenth century to the fourteenth century, recently described by Geoff Wade as an 'Early Age of Commerce', Butuan in northern Mindanao boasted local specialized craft production that included weaving, boat-building, and metal-working, most notably gold-smithing. Appearing in the *Song Hui-yao Ji-gao*, Chinese sources of the Song dynasty (eleventh – mid-thirteenth centuries), Butuan seemed to have been a dependency of Cambodia (Wade 2009a:221–265). The source identifies an envoy from Butuan who had travelled to the Imperial Court in the year 1003, the first of several recorded tribute missions. Along with gifts of cloves from the Moluccas, white camphor from Borneo, pearls from Sulu, sandalwood possibly from Timor, tortoise shell, and red parrots from New Guinea, the envoy presented a memorial inscription

22 Reid (1988), I, (2000c); Wade (2009a). Victor Lieberman (1995), however, has contended that these 'Ages of Commerce' models are not applicable to the histories of mainland Southeast Asia.

23 Flecker (2004–2005:2). For an excellent overview of commodities and cultural exchange, see Scott and Guy (1995).

engraved on a tablet of gold (Wade 1993:84–85; Guy 2011a:167). Recovery of a diverse and plentiful range of high-fired ceramic tradewares indicated the participation of northeast Mindanao within a Philippine-Borneo-Celebes network and, more broadly, its place in Song era Chinese, Southeast Asian, and Middle Eastern trade patterns.[24]

Conclusion

In this chapter we have noted the great range of Southeast Asian *naturalia* that was traded from the third century BCE onwards: pearls, coral, tortoise-shell, civet, kingfishers' feathers, live birds with brilliant plumage and the ability to talk; aromatic woods, resins, ivory and rhino horn. Such things have been treated here not merely as valuable mercantile commodities sought out by the wealthy and powerful, but also as objects of intrinsic interest to the romantic, to the aficionado, to the avaricious, and the medically minded. In the European Renaissance, early globalization, commerce, and imperial ambition produced very extensive shopping lists of *naturalia* and also stimulated the activity of collecting, the artistic depiction of the world's natural diversity and the invention of natural history as a discipline (Ogilvie 2006). Trade and commerce were at the heart of collecting activities. In our period of study, the demand for natural objects emerged out of certain conditions – economic exchanges, networked land and sea routes, cultural interactions and the movement of peoples over the *longue durée*.

Knowledge about Southeast Asia, as we have seen, was constructed through the production of maps and books that devised meticulous classificatory systems which were useful for trade, but which also assembled medical, scientific, geographic, and ethnographic information. Some travellers' told tales of strange creatures and fantastical places. Southeast Asian lands and peoples came to be regarded as distinctive, even alien, with an eminently commercial and exploitable natural environment. Finally, it is clear that the desire to acquire *naturalia* transcends time and place. Many reasons have driven the search for, and possession of, the rare, the strange, the exotic, the precious, and the sensual. The impetus could come from lofty aims – the desire to discover medical panaceas, the spirit of scientific inquiry, and the reverence for objects

24 Most abundant were Chinese trade ceramics, the oldest being Yue and Yue-type wares dating from the Five Dynasties period (907–960), but also included were Thai, Vietnamese and ninth-century Middle Eastern objects. Brown (1989:79).

regarded as sacral and ritualistic. Humans have been moved by the whims of cultural taste, aestheticism, curiosity, fascination, ambition and vanity. These impulses long preceded the European Renaissance, and their history informs our understanding of humankind's interactions with nature as we continue to confront environmental destruction and disappearing bio-diversity on a global scale.

CHAPTER 8

Ages of Commerce in Southeast Asian History

David Henley

Fernand Braudel classically proposed three interlocking historiographic time frames: the 'history of events', the history of cycles, and the *longue durée* or long duration (see this volume, Chapter 1). As an example of cyclic historical change, Braudel (1960:5) cited the fifty-year cycle of rising and falling prices identified in European history by the economist Nikolai Kondratiev. In Southeast Asian history, as Boomgaard (2002) has shown, the Kondratiev cycle is only of late and limited significance. In recent literature, however, another, much longer and more important economic cycle has emerged in the historiography of Southeast Asia: an epic succession of centuries-long 'ages of commerce', separated by trade depressions lasting many decades. Although approximately cyclic, the chronological scale of this phenomenon seems to put it firmly in the realm of Braudel's *longue durée*, as do the profound political and cultural consequences which some writers have ascribed to it. The present chapter provides a brief critical survey of the development, implications, and limitations of the Age of Commerce paradigm.

Not so long ago, the history of commerce in Southeast Asia was for most purposes equated with the history of colonialism. Money and trade were seen essentially as one component of a great Western assault – economic, political and cultural – on 'traditional' Southeast Asian societies that began in the sixteenth or seventeenth century, and led ultimately to the nationalist reaction of the twentieth (Boeke 1953:67–86; D. Hall 1955:649–671). It was Europe which, with a little help from Chinese 'middlemen', had intruded upon and broken open Southeast Asia's age-old self-sufficiency. In this view, epitomized by B.J.O. Schrieke's memorable declaration (1955–57, II:100) that 'the Java of around 1700 AD was in reality the same as the Java of around 700 AD', a millennium of virtual stasis – in economic as in other spheres – was brought to a late, dramatic end by the accelerating rise of Western power in the eighteenth and nineteenth centuries.

Of course there was always a memory of the fact that the very earliest European interventions in the region, in the sixteenth and seventeenth centuries, had consisted of attempts either to extract parasitic profits from pre-existing intra-Asian trade systems (as in the case of the Portuguese occupation of Malacca), or to destroy and supplant them (as in the case of the VOC assault on the Moluccas). But scholars who had looked at this period tended to characterize the precolonial sea trade of Southeast Asia as a primitive 'peddling' of luxury goods from port to port, having 'nothing in common with "bourgeois" commercial forms of trade' (Van Leur 1960:54–55). Compared with nineteenth-century developments like the ruthless state-orchestrated commercialization of the Dutch Cultivation System, the free-trade fanaticism of British Singapore, the 'opening up' of Siam under British pressure following the Bowring Treaty, and the great military *mise en valeur* of French Indochina, the early episodes in which Asian commerce was restricted or suppressed by European arms seemed inconsequential counter-currents within a larger history whereby the expansion of trade was overwhelmingly associated with European expansion.

A Precolonial Age of Commerce, 1400–1650

It was Anthony Reid (1988–93, 1990a, 1990b), albeit building on earlier revisionist work on precolonial trade by Meilink-Roelofsz (1962) and others, who at the end of the 1980s initiated a paradigm shift. Reid presented evidence that a serious economic (and demographic) as well as political crisis took place throughout island Southeast Asia during the period of VOC expansion in the seventeenth century, and identified that century as marking the *end* – not the beginning – of a Southeast Asian Age of Commerce.

During the period 1405–1630, Reid argued, a long boom in pepper and spice exports to Europe (first via Arabia and the Mediterranean, later via Portugal) and to China led a general expansion of commerce throughout Southeast Asia, triggering in turn a series of new developments in the cultural, social and political spheres. These included mass conversions to Islam in Indonesia, the establishment of the modern Theravada Buddhist orthodoxy in mainland Southeast Asia, the growth of trading cities like Ayutthaya, Malacca and Makassar, and the formation around them of new centralized states financed by trade wealth. Toward the end of this Age of Commerce, the general movement away from local economic autarchy toward commercial exchange and specialization was supported by the emergence of a new pan-regional currency in the form of Spanish American silver imported via the Philippines.

In the mid-seventeenth century, according to Reid, the trade boom ended due to the military conquests and spice monopolies of the VOC, together with the threat of further Dutch aggression against polities encouraging non-VOC commercial activity. Among the earliest signs of the crisis was the complete disappearance by 1650 of the very large Southeast Asian sailing vessels, efficient but vulnerable to naval attack, which had previously carried much of the region's commerce. Starved of trade income, those major Indonesian states which avoided Dutch conquest (Aceh, Java) declined and fragmented, while on the mainland, Burma and Siam retreated defensively from commerce with Europeans. Other factors contributing to the trade slump included internal political problems within the increasingly absolutist indigenous polities, the rulers of which tried, like the VOC, to restrict commerce by military means in order to deny economic resources to (actual or potential) rivals.

An Early Age of Commerce, 900–1300

Recently Wade (2009a), inspired by Reid, has posited another, earlier Southeast Asian Age of Commerce beginning five centuries before Reid's and lasting until about 1300. In many ways, Wade argues, developments during this 'Early Age of Commerce' paralleled and anticipated those of its later counterpart: a boom in maritime trade, the emergence of new ports and urban centres (notably Srivijaya in Sumatra), a movement of existing administrative capitals toward the coast, commercialization, monetization, population expansion, and also religious change, including the first beginnings of the Islamic and Theravada Buddhist expansions which were to bear full fruit during Reid's period.

Triggered by outward-looking trade policies under the Song and Yuan dynasties in China, and by a Chinese 'medieval economic revolution' involving the creation of new forms of money and credit, the maritime trade boom of Wade's Early Age of Commerce was based on the export from Southeast Asia to China of pepper, safflower, fine spices, and forest and marine products, and on the import to Southeast Asia of Chinese ceramics and metals. At the same time Indian Ocean commerce also burgeoned under the auspices of two long-distance maritime trading systems: an Islamic merchant network extending from Arabia to the ports of China, and an almost equally extensive Tamil network emanating from the Chola kingdom on the Coromandel coast of India. The commercial globalization of the period has been highlighted by the recent discovery and excavation in Indonesian waters of five contemporary shipwrecks (including one of a vessel built in India or the Middle East), complete with cargoes including Chinese iron and ceramics, Indian bronzes, and Arabian glass.

Wade links this early globalization with evidence for general economic growth and political change in Vietnam, Champa, and especially Java. Here he draws heavily on earlier work by Jan Wisseman Christie (1993, 1996, 1998, 1999) on Java's role in what she explicitly calls 'the Asian sea trade boom of the tenth to thirteenth centuries'. Java, Wisseman Christie shows, became a sophisticated market economy in this period, attracting many foreign merchant communities and minting its own coins in gold, silver and electrum as well as importing large quantities of copper cash from China. In 929 CE, she also notes, the political centre of Mataram, Java's dominant state, moved from the landlocked interior of Central Java to a coastal location on the Brantas delta in East Java, where the great fourteenth-century 'empire' of Majapahit would later have its capital. By the time of Majapahit, however, the Early Age of Commerce as defined by Wade had already come to an end around 1300 due to 'stagnation in southeast Asian maritime trade, possibly as a result of the years of warfare the Mongols waged against the southern Song in China, the Yuan efforts to tightly manage maritime trade and the major Yuan military missions launched in the maritime realm' (Wade 2009a:264).

Even Earlier Ages of Commerce?

Besides describing the Early Age of Commerce *avant la lettre*, in 1998 Wisseman Christie began to sketch the outline of an even earlier Asian sea trade boom, beginning no later than the seventh century and ending in the ninth, when China 'turned inwards' toward the end of the Tang dynasty. She also pointed to 'hints in the archaeological and early historical records' that ultimately the great trade cycle 'may have had its roots in the later prehistoric period', when Southeast Asia was known in India as a source of gold and precious metals (Wisseman Christie 1998:344).

It is interesting to note that the very early trade boom – if that is what it was – which ended in the ninth century coincided with the original cultural Indianization of Southeast Asia, with its Sanskrit (not yet Javanese or Khmer) inscriptions and its 'prototypical' Hindu and Buddhist temples such as those of the Dieng Plateau. This was the era of what Sheldon Pollock has called the 'Sanskrit cosmopolis', when 'Sanskrit literary culture spread across most of southern Asia from Afghanistan to Java', and when the religious, political and artistic ideals of elites throughout that cosmopolis became so similar that 'in the first millennium it makes hardly more sense to distinguish between South and Southeast Asia than between north India and south India' (Pollock 2006:1, 16).

For decades it has been customary to assert that the eastward spread of Indian cultural patterns was the result of Southeast Asian rather than Indian initiative, and did not reflect any kind of 'colonization' by South Asian people, as many once believed (Majumdar 1963). The artistic forms and Sanskrit inscriptions of Indianized Southeast Asia are certainly ambiguous in this respect, being open to Pollock's interpretation as evidence of decentralized cosmopolitanism rather than migrant-borne cultural transmission. Nevertheless we know from the epigraphic record that during the first millennium there were in Southeast Asia diverse communities of Indian traders – one of which even left, in Sumatra and the Malay Peninsula, a series of inscriptions in its own vernacular language, Tamil, created *in situ* over a huge span of time between the third or fourth century and the thirteenth (Guy 2011b).

More importantly, evidence is mounting from genetic research that early immigration from India played a much more important role in the history of Southeast Asia than it has lately been conventional to assume. According to a recent analysis of DNA samples from more than 500 Balinese men, 'approximately 12% of the Balinese paternal gene pool' is of Indian origin (Karafet et al. 2005:94).

> The discovery that about 12% of Balinese haplogroups are of relatively recent Indian origin and that these haplogroups are rare or absent in neighboring Indonesian islands should prompt a thorough re-examination of the archeological evidence for contact between India and Bali. The genetic evidence we have presented suggests that the magnitude of such trade and other cultural contacts between India and Bali was much greater than has hitherto been imagined.
>
> KARAFET ET AL. 2005:111

A further surprise is that estimates of the antiquity of the Indian genetic input into the Balinese population, derived from the same data, range from 2,600 to 3,100 years, putting it well before the time of the Sanskrit cosmopolis. The discovery of the tooth of an Indian visitor or immigrant in a Balinese archaeological context from before the beginning of the Common Era confirms the prehistoric origins of direct contact between South and Southeast Asian populations (Lansing et al. 2004).

Colonial and Postcolonial Ages of Commerce in Comparative Perspective

The identification of at least three major cycles of commercial expansion and contraction in precolonial times places the history of the colonial and postcolonial periods in a radically new light. The great trade boom of the nineteenth and early twentieth centuries, together with the period of financial crisis, war, and political turmoil that followed it after 1930, becomes just one in a long succession of basically comparable cycles, and the renewed regional economic growth and globalization which began around 1970 fits into the same established pattern. Even the cultural aspects of these two most recent cycles can be seen as echoing those of their predecessors – Southeast Asia's current integration into an Anglophone global cosmopolis, for instance, parallelling in some ways the participation of its elites in the Sanskrit cosmopolis of the first millennium.

It is striking that when the successive Ages of Commerce are tabulated chronologically alongside the intervening crises (Table 8.1), Southeast Asia's history is seen to consist predominantly of the former. Taken together, the crises of the ninth, fourteenth, (long) seventeenth and (short) twentieth centuries occupy just over one quarter of the last 1400 years. The other three quarters fall within periods of relative commercial openness and growth, indicating that in the very *longue durée*, the Southeast Asian past is dominated by a secular trend of commercialization.

Nevertheless it would be misleading to exaggerate either the cumulative nature of this process, or the mutual comparability of the various commercializing phases. If the nineteenth century was not Southeast Asia's first Age of Commerce, in terms of sheer scale and value of trade it still represented an unprecedented change. In 1998 Reid and a number of colleagues published a slim yet monumental compilation of statistics on exports of four major Southeast Asian commodities – cloves, pepper, coffee, and sugar – to other parts of the world from the fourteenth century to the twentieth. When the total value (corrected for inflation) of all four is plotted graphically by decade (Figure 8.1), it is clear that by far the fastest export growth took place during the nineteenth century, and that compared with the level of exports reached by

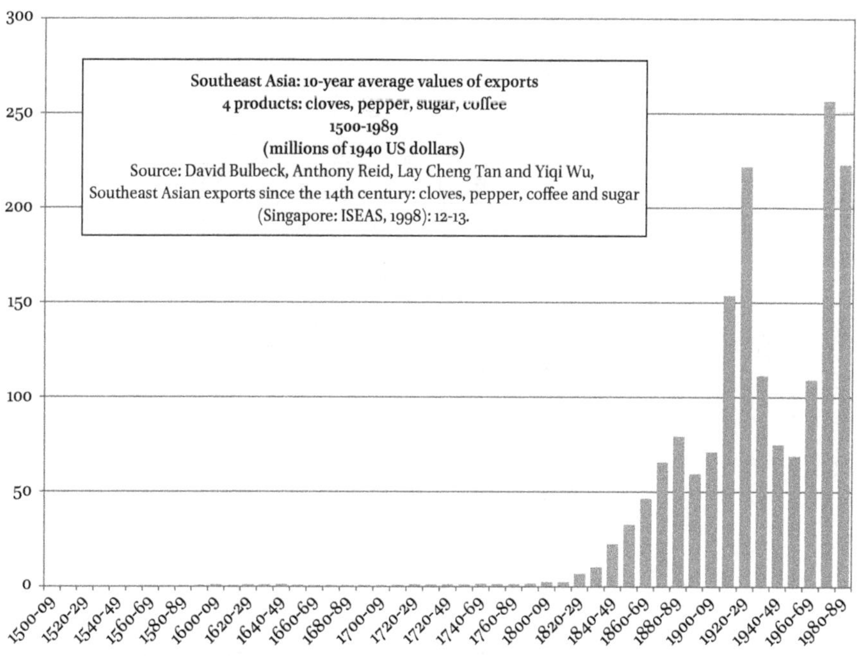

FIGURE 8.1 *Southeast Asian exports, 1500–1990*

TABLE 8.1 *Ages of Commerce in Southeast Asian history*

		Politics	Economy	Culture
First Asian Trade Boom	?-800	early Indianized polities in SE Asia (Funan, Srivijaya)	export of forest products and spices; import of Chinese ceramics; China-India transit trade via Straits and Kra	Sanskrit cosmopolis; Hindu-Buddhist religious culture
9th-century crisis	800–900	breakdown of Tang state disrupts Chinese trade with and through SE Asia		
Early Age of Commerce	900–1300	outward-looking trade policies in Song China; Muslim shipping spans Indian Ocean and South China Sea; Tamil trading networks supported by Chola dynasty; SE Asian states seaport-oriented (Srivijaya, Mataram from 929)	export of forest products (ebony, betelnut, beeswax, aromatic and medicinal plants), dyestuffs, pepper; import of Chinese ceramics and iron; monetization (indigenous silver and gold coinage, Chinese copper cash)	Classical (Indic) Java and Angkor; introduction of Theravada Buddhism and Islam
14th-century crisis	1300–1400	war and instability in Yuan (Mongol) China; sea trade redirected to overland Silk Road; collapse of Cholas in India; Srivijaya disintegrates; collapse of Angkor; conflict and fragmentation in Burma		

<table>
<tr><th></th><th></th><th>Politics</th><th>Economy</th><th>Culture</th></tr>
<tr><td>Age of Commerce</td><td>1400–1650</td><td>Ming China looks outward; rise of Malacca as Straits transit port; trade-financed political centralization in SE Asia (Java, Aceh, Burma, Siam); first European contacts and conquests (Malacca, Moluccas, Java, Philippines)</td><td>cash cropping (cloves, nutmeg, pepper); apogee of indigenous shipping (SE Asian junk); European commerce; Indian cloth imports; import of silver coinage; urbanization</td><td>Islamization (Indonesia); Christianization (Philippines); modern Theravada Buddhist orthodoxy and institutions established on mainland</td></tr>
<tr><td>long 17th-century crisis</td><td>1650–1780</td><td colspan="3">Dutch monopolies constrict maritime trade; internal conflict in Indonesian states (Java, Aceh); defensive retreat from commerce on mainland (Siam); cultural retraditionalization</td></tr>
<tr><td>Colonial Age of Commerce</td><td>1780–1930</td><td>Free trade at sea restored; extensive colonial conquests; forced commercialization in Java</td><td>cash cropping (bulk crops); mass labour immigration; industrial imports; technical irrigation</td><td>movements toward religious orthodoxy; (from 1900) Western education, nationalism</td></tr>
<tr><td>20th-century crisis</td><td>1930–1970</td><td colspan="3">global financial crisis (1930s), war, revolution, economic nationalism</td></tr>
<tr><td>(post-) modern globalization</td><td>1970–present</td><td>counter-revolutionary developmental states; liberalizing trade regimes</td><td>Green Revolution; export-oriented industrialization</td><td>global cosmopolis; Islamic revival</td></tr>
</table>

1900, those achieved during the last precolonial Age of Commerce, even at its peak in the 1640s, shrink into insignificance. Since population also grew in the nineteenth century, the contrast with respect to the four selected commodities would be somewhat less dramatic if their value was calculated on a per capita basis. On the other hand the difference in absolute value would be even greater if important new export products of the late colonial period, such as copra, rubber and oil, were included in the statistics.

Limitations of the Paradigm

One disadvantage of the Ages of Commerce paradigm, then, is its tendency to mask the novelty, at least in a quantitative sense, of what happened in the nineteenth century and after. Reid (1998b:15) highlights the fact that within the colonial period, the rate of export growth was higher in the period 1780–1870, when Southeast Asia's export crops were still produced almost exclusively by small farmers, than in the decades of 'high colonialism' after 1870, when big European-owned plantations became significant. In so far as the trade boom of this period was underpinned by the ability of Southeast Asian peasants to seize commercial opportunities, there was indeed continuity here with earlier Ages of Commerce. Nevertheless, the colonial era in Southeast Asia was clearly a revolutionary one in economic terms. Besides production for export on a scale much more comparable with the twenty-first century than with anything prior the nineteenth, it also saw movements of capital and labour to and within Southeast Asia on a scale which, after the crisis of 1930, was not to be seen again for more than 50 years (Amrith 2011:29–49, 151; Lindblad 1998:24).

Reid (1998b:15) dates the onset of the trade recovery, following the crisis of the long seventeenth century, to the 'collapse of Company control' and the beginning of British naval ascendancy around 1780. This is consistent with his portrayal of the previous Age of Commerce as having been possible thanks to an environment of free trade, and having ended largely because of the VOC's endeavour to monopolize the sea lanes and control the production of the most valuable export goods. However it is worth noting here that during the period of very rapid growth in Southeast Asian exports in the middle of the nineteenth century, a large proportion of that growth was accounted for not by free peasant enterprise, but by the compulsory (albeit not necessarily unprofitable for the cultivator) production of sugar and coffee by an unfree peasantry under the Cultivation System in Dutch Java (Reid 1998b:13, 123–124, 152–153).

A more fundamental criticism of the paradigm is that as far as the period prior to the nineteenth century is concerned, the propositions advanced in the

Ages of Commerce literature are based on weak statistical foundations. Reid and his colleagues do an admirable job of bringing together what scattered quantitative data do exist on Southeast Asian spice exports, and silver and cloth imports, in the early modern period. But is difficult to avoid the suspicion that the figures they synthesize still represent only a fragment of reality, and that the trends they infer often reflect changes in the coverage and accuracy of their sources rather than real changes in economic conditions. Some of the population figures cited by Reid to support the idea of a seventeenth-century demographic crisis, consisting of crude household counts compiled for taxation purposes, are almost certainly so unreliable as to be useless for historical purposes (Henley 2005a:115–116).

Even allowing for statistical understatement, the limited quantity and value of precolonial long-distance trade remains disconcerting given its alleged historical significance. The further we go back into the past, not surprisingly, the more striking this discrepancy becomes. The best-known objects of (probable) South Asian origin found in Southeast Asian archaeological sites dating from the eve of the Sanskrit cosmopolis are fragments of a type of pottery known as 'rouletted wares', sometimes used as a marker of Indian Ocean trade. Manguin (2011:xxi) notes that if all the shards of this ware ever found in Southeast Asia, representing some four centuries of commerce, were put together, they would make up 'enough whole dishes to set tables for only five dozen people'. While far greater numbers must of course have disappeared in the intervening 2,000 years, such a statistic still seems more in line with the old anthropological paradigm of 'prestige goods exchange' than with any *economic* model of Indian Ocean commerce.

Although the long-distance trade of the medieval and early modern Ages of Commerce was undoubtedly more intensive, it remains hard to disagree with Van Leur's well-known assessment that most of the commodities involved were still 'splendid and trifling': incense, dyestuffs, spices, porcelain, silver. Could their transport and exchange really have had the profound historical implications claimed for them: the rise and fall of states, economies, populations, religions? And if commerce played such a central role in cultural change, then how to explain the feeble cultural influence of China on Southeast Asia (except Vietnam) over a millennium of Chinese trade and immigration? Or the fact that the modern religious institutions of Burma, Thailand, Laos and Cambodia have their origins in Sri Lanka, a country not known to have traded intensively with Southeast Asia? These puzzles raise fundamental questions of causation. Did the trade cycle really cause the other changes associated with it? If so, what drove the trade cycle? If not, why the apparent synchronicity? Was there some other motor that powered them all?

Lieberman's Critique

Although he claimed it was also applicable to parts of the mainland, particularly Thailand, Reid's Age of Commerce thesis was developed mainly in relation to the maritime realm of island Southeast Asia (present-day Indonesia, Malaysia, and to a lesser extent the Philippines), which he knew best and for which the best early data existed. It was soon to be contested by the doyen of mainland Southeast Asian historical studies, Reid's counterpart Victor Lieberman (1995, 2003, 2009, 2010).

Lieberman found little evidence in Burma, Thailand or Vietnam for Reid's seventeenth-century crisis. Here the decline in European trade during that century was compensated by growing Chinese commerce, and the major states became stronger rather than weaker between 1650 and 1750 (Lieberman 1995:801–803). At least in terms of political disintegration, however, three other major crises were clearly discernible on the mainland. These occurred in the fourteenth century, which saw the demise of the classical empires of Angkor and Pagan; in the late sixteenth, when the main kingdoms in Burma and Thailand both briefly collapsed; and in the mid-eighteenth, when conflict and revolt (including the Tay Son rebellion in Vietnam) overtook the whole peninsula. Of the three mainland crises only one, that of the fourteenth century, clearly corresponded to events in the islands, where it coincided with the fall of Srivijaya – although Majapahit, further east, was temporarily able to benefit from the resulting power vacuum by strengthening its influence over the Spice Islands (Wisseman Christie 1993:4–5). All three, on the other hand, seem to have surprising echoes in much more distant parts of Eurasia, including France, Russia, and to some extent Japan (Lieberman 2003:3, 4; 2009:58–62).

International trade cycles, Lieberman argues, cannot in themselves explain these patterns. This is not only because mainland Southeast Asia was for most of its history much less commercialized than the islands – astonishingly, the great medieval empire of Angkor, in sharp contrast to contemporary Java, was still 'a society without money' (Lustig 2009:88) – but also because commercial links with Europe and Japan were simply too tenuous, at most of the periods in question, to provide a convincing explanation for the observed cyclic coordination.

Lieberman's own theory, reduced to essentials, is as follows. First, big polities throughout Eurasia were individually subject to an endogenous cycle of expansion and decline caused in the ecological sphere by a cycle of agricultural intensification, population growth, and overpopulation; and in the political sphere by a cycle of territorial overexpansion and fragmentation, combined perhaps with 'an inherent tendency toward dynastic cyclicity', whereby 'an initially successful dynasty was obliged to tolerate the accumulation of resources by political or religious elites on whom it depended', but who eventually usurped its power (Lieberman 2003:369). The reasons why these cycles became more or less synchronized across the continent, secondly, included intellectual and technological diffusion, disease dynamics, and Central Asian politics, as well as international commerce (Lieberman 2009:77–92). Perhaps the most important reason, however, was climatic change.

Nature as Pacemaker?

Not all environmental forces shape human history only as steady constraints in the *longue durée*; some are themselves subject to change over historical time. Reid (1990b:654–656) already suggested that climatic variation, in the form of a series of very dry years in the 1650s and 60s – confirmed by evidence from growth rings in teak trees from Java – exacerbated the crisis of the seventeenth century by causing poor harvests, famine, and epidemics. Lieberman (2003:239–240) took this further by drawing on work by physical geographer David Godley to argue that the collapse of Angkor and Pagan in the fourteenth century was caused partly by a 'general desication of Indochina' (Godley 1997:142), coinciding with the end of the 'medieval warm period' in Europe, when England produced wine and Greenland was settled by Vikings. Recent dendrochronological research on teak and other deciduous trees from Burma, Thailand and Vietnam confirms that periods of repeated drought and climatic instability coincided not only with the fourteenth-century crisis (Buckley et al. 2010), but also with the political crises of the mid-eighteenth century (D'Arrigo et al. 2011:4).

In Table 8.1, the rhythm of the Ages of Commerce and the intervening crises appears to be set mainly by political events (such as dynastic changes and wars) originating outside Southeast Asia in China, India, and (in the case of the seventeenth-century VOC intervention) the West. But advances in climate history may yet show that most of those events are themselves related, via ecological, demographic, and economic intermediary mechanisms, to an underlying rhythm of climatic change that acts as the global pacemaker of human history.

Besides climate change, other important environmental events influencing the course of human history include the volcanic eruptions, earthquakes and tsunamis which Indonesians in particular are once more all too familiar with in the early twenty-first century. Belonging in a sense more to Braudel's *histoire événementielle* than to the *longue durée*, these can in fact be more sudden and intermittent in their effects than any war or economic crisis, yet at the same time so violently destructive that their after-effects may last for decades, or even longer. Recently, Reid (2012) has used a combination of geomorphological evidence, uncovered by research stimulated by the 2004 Aceh tsunami, and historical sources to suggest that hitherto unsuspected tsunami disasters were involved in both the seventeenth- and fourteenth-century economic and political crises. The idea that such a critical watershed in Indonesian history as the transition from Indianization to Islamization, which transition is arguably one dimension of the fourteenth-century crisis, can be explained partly by a combination of climate change and tsunamis is for most people – specialists included – a new, radical, and intriguing one. Along with experts in historical genetics, environmental scientists, using radiocarbon-dated material from coral reefs and beach deposits to reconstruct seismic and tsunami events, and data from tree rings to reconstruct climate change, are now at the forefront of research into the history of Southeast Asia in the *longue durée*.

Conclusion

The aim of this chapter has been to provide a brief critical survey of the development, implications, and limitations of the Age of Commerce paradigm for the study of Southeast Asian history. The idea of an Age of Commerce, initially proposed by Anthony Reid to characterize the period 1400–1680, has been a fertile one, prompting both a search for parallels with other periods and an extensive critique of Reid's chronology and conclusions, particularly in relation to mainland Southeast Asia. The economic and demographic evidence informing this debate is often surprisingly weak, and those statistics which do exist suggest that quantitatively speaking the long-distance commerce of Reid's and earlier periods, whether measured by volume or by value, never amounted to more than a tiny fraction of what Southeast Asia was to export and import in the nineteenth and twentieth centuries. Nevertheless, the evolving debate over the Age of Commerce paradigm has had many benefits for our understanding of Southeast Asia. It has, for instance, confirmed that openness to commerce and globalization has been characteristic of the region, and particularly of maritime Southeast Asia, throughout most of its history. It has helped to undermine oversimplified dichotomies between colonial and precolonial situations which have often led the impact of Europe on Southeast Asia to be exaggerated or misunderstood. It has helped to draw attention to the importance of environmental forces, including climate change and seismic events, in shaping Southeast Asian history. And it has contributed to the general, pervasive awareness of the *longue durée* and its patterns which is reflected in the work of influential contemporary Southeast Asianists such as Peter Boomgaard.

CHAPTER 9

Pursuing the Invisible

Makassar, City and Systems[1]

Heather Sutherland

An archaeologist, trying to reconstruct an entire skeleton from a thigh bone and a tooth, works with tested structural models; the final result clearly distinguishes between the actual (bone) and hypothetical (plaster or resin). In the pure sciences the interlocking of theory and data ensures that knowledge limitations are explained. But historians are often innocent excavators, following whatever sources happen to survive, filling gaps with assumptions or, more helpfully, with clearly expressed best-guesses. Such evasions are difficult to avoid when narratives cover large swathes of the globe or many centuries. Despite sparse and unevenly distributed information, such histories must encompass very different local situations, and the need for a comprehensive story-line encourages generalization and the inflation of specific case studies to represent much wider topics. Themes and comparisons may be explained, but methodology – the sources and how they are used – is often taken for granted. It has been suggested, for example, that much world history is anecdotal, 'not really having a principle of selectivity. As a result, courses in the subject are often a pastiche, made up of bits and pieces of research into various parts of the world and stitched together in hit or miss fashion' (Mazlish 1998, 2001:4). On the other hand, in more focused, but no less ambitious, publications explicit theory and a comparative framework can serve to focus exhaustive research and discipline creative imaginations, as Peter Boomgaard has so ably demonstrated. Often, however, models are implicit.

Long-term narratives frequently begin with a base-line in order to track linear, evolutionary change. Here much depends on the validity of the point of departure; if this is misconceived subsequent analysis is distorted. An alternative strategy takes the present as model, working back to locate underlying continuities and fundamental structures.[2] This can also be risky, creating self-fulfilling prophecies. Moreover, whatever method is used, source material will be limited. Significant dimensions of the subject remain unknown, but may be

1 I am grateful to Leonard Andaya, David Bulbeck and Campbell MacKnight for their helpful answers to questions concerning the early history of Makassar.

2 A classic is Robert D. Putnam, Leonardi, and Raffaella Y. Nanetti (1992).

highly relevant. I would suggest that this should be acknowledged, and that the use of explicit models can enable us to sense the contours of the whole despite missing information.

This essay uses the history of one particular place to discuss the ways in which contextualization within a systems framework can improve our modelling of the past. The focus is upon Makassar, a port-town on the island of Sulawesi (Celebes) in East Indonesia, a region of scattered archipelagos extending over waters as wide as those of the entire Mediterranean.[3] The first section gives a short introduction to the region, before considering how a simple systems perspective can help locate hidden structural relationships, if only by inference. In this light I then re-evaluate Makassar's role as an illustrative example in Anthony Reid's path-breaking study of the 'Age of Commerce', a standard base-line in many Southeast Asian histories.[4] The third section introduces retrospective accounts of the town's locating in regional political and economic patterns, while the conclusion summarizes how an awareness of encompassing systems could better balance these narratives.

The Region

The eastern archipelagos of Southeast Asia were the peripheries of a periphery, characterized by weak formal institutions, but a rich tradition of trans-regional exchange. Unfortunately, information on Eastern Indonesia is sparse. Chinese and Arab sources are important for the very early period, but for the centuries between 1500 and 1800 it is the rather bureaucratic Portuguese, Dutch and English trading companies that provide rich documentation, although it is focused on long distance commerce in specific products and on elite politics. Histories of the region since the fifteenth century traditionally stress its 'open' nature, created by geography and reflected in long-distance trade and the absorption of new religions. The most persistent, and pernicious, assumption has been that innovation was always driven by external forces (Sutherland 1995). Southeast Asians were seen as being on the fringe, passive recipients of influence from the stronger civilizations of India, China and the West. Such marginalization once implied insignificance, but the periphery is now a fashionable place, as everywhere is 'open'. Globalization has weakened the idea of centrality, and the view from the edge now seems particularly clear-sighted.

3 If the east coasts of Kalimantan (Borneo) and Spain were aligned, West New Guinea would lie on the Syrian coast.

4 Anthony Reid (1993).

Braudel's Mediterranean has provided a tempting archetype for maritime Southeast Asia, where economic, cultural and social exchange created solidarities that transcended political boundaries (Sutherland 2003). But Southeast Asia was very different and Eastern Indonesia even more so. Firstly, the enclosed maritime basin of the Middle Sea – Braudel's basic geography – contrasts with Eastern Indonesia's many lesser seas (Sulu, Celebes, Maluku, Banda, Flores, Arafura, Java) which connected the great maritime routes linking the Indian and Pacific Oceans with the South China Sea. Secondly, when compared with Java or the Southeast Asian mainland, let alone most Mediterranean hinterlands, these islands were extremely sparsely populated. High levels of insecurity and mobility among shifting cultivators and sea-farers meant that communities were shallowly rooted and quick to shift their homes or allegiance. Trading and raiding were intertwined, local conflicts (often related to headhunting) were endemic, and punitive expeditions were central to statecraft. Thirdly, states, settlements and traders were linked by personal networks; institutions were informal and contingent. Fourthly, the most prized products shipped across these waters came not from agriculture or industry, but from foraging and fishing. Productivity was limited by poor tropical soils, limited flat land, scarcity of good harbours and, in the south-west, unreliable rainfall. The only reason that long distance shipping came to the region at all was the fortuitous existence of a few very valuable commodities growing in very limited areas.[5]

Supplies of regional products were mobilized by coercion as least as much as by commerce. East Indonesian regional maritime powers had much smaller populations and resource-bases than those in the west; almost all depended on imported foodstuffs.[6] Successful states combined a reasonably reliable food supply with access to imports. Those rare regions with land suitable for irrigated rice fields, such as Southwest Sulawesi, were particularly lucky. But even there polities were typically kin-based federations, sanctioned by divine origin, and characterized by a rigid social hierarchy, the competitive mobilization of personal followings, conflict and constantly shifting patterns of alliance and hostility. Family ties and brute strength were crucial in determining access to power. Migration overseas was common.[7] However, most of the eastern

5 Cloves only grew around Ambon, nutmeg at Banda, sandalwood at Timor and, Sumba while the dye-producing sappanwood was mainly from Sumbawa.

6 The most notable being Kutai on the east and Banjarmasin on the southeast of Kalimantan (Borneo); Ternate and Tidore in the Moluccas; South Sulawesi's Makassar, Bone etc.; and, on the fringe, the Balinese kingdoms and the lesser state of Bima on Sumbawa. The Southern Philippines Sulu and Mindanao Sultanates were very similar.

7 L. Andaya (1975); Sutherland (1983); Anderson (2003); Bulbeck (1996); Caldwell (1995); Cummings (2002).

archipelagos were without any over-arching authority at all. Chiefs seeking imported prestige goods and weapons competed to attract passing vessels en route to regional entrepot where valued commodities could be found. Location close to a major trade route, such as those connecting the Straits of Melaka, southern China, or island sources of spices or fragrant woods, was thus essential. It is in this context that Makassar should be placed.

The long-term history of such maritime zones can be modelled as layered, de-centred structures of inter-locking but open and non-territorial subsystems. Systems theory focuses on 'understanding the interaction of the various elements of a complex system: it [...] concentrates on modelling complex wholes rather than breaking large problems into smaller problems for individual solution' (Manning 2003:290). Relationships within systems are generally many-stranded (economic, social, political) and multiple, being neither exclusive nor fixed. Locating the centres or nodes of open systems, with their inflows and outflows, is easier than fixing their boundaries.[8] Feedback within and between systems and subsequent adaptive behaviour ensures that 'circular causal chains help to maintain a steady state within the system'.[9] A breakdown in adaptation heralds structural change; a steady state is not a permanent equilibrium. The popular concept of 'network' is similar in its emphasis on un-institutionalized linkages, but lacks systems theory's explicit focus on inputs, outputs and adaptation. The idea of adaptation is particularly useful as it avoids the dichotomy of continuity and change, and modifies the emphasis on the tension between fragmentation and integration as the overriding dialectic in Southeast Asian politics and commerce (Reid 1993, II; Lieberman 2003, 2009).

Port-towns offer an identifiable, but complex, point of entry. Half a century ago Brian Berry (1964:160–161) argued that 'cities may be considered as systems – entities comprising interactive, inter-dependent parts...[S]ets of cities also constitute systems. For systems of cities the most immediate environment is the socio-economy of which they are a part'.[10] The main source for the following section is the VOC or Dutch East India Company (est.1602) archive, but the inspiration comes from Berry.

8 Chase-Dunn and Hall (1997:16–20); Kratoska et al.:(2005). Boundaries would logically fall at points where the relationships contained within the line are of more significance than those that cross it (Subrahmanyam 1997:740).

9 Manning (2003:292). Analysis of adaptive behaviour addresses questions of structure and agency.

10 See also Timberlake (1985); Dick and Rimmer (2003).

Looking Forward: The 'Age of Commerce' (1450–1680) as Base-Line

In the first volume (1988) of his two volume Braudelian study Anthony Reid (1988, I:6) introduced two key ideas: that of 'cosmopolitan port cities', and that of a mid- seventeenth century 'trade revolution' of rapid decline 'as indigenous cities withdrew from international commerce, or were defeated by the Dutch commercial monopoly'. By the time the second book was published (1993) this 'trade revolution' had become a more generalized 'crisis' (Reid 1993, II:23). Specifically, Reid (1993, II:303) observed that 'Makassar and Banten, conquered by the Dutch in 1666–1669 and 1682, respectively, lost their political and economic *raison dêtre,* and declined to no more than a quarter of their previous populations'. He had already stated that 'the destruction of the cosmopolitan trading cities of Banten and Makassar was immensely important for the evolution of the societies of which they were the heart...they became embittered backwaters' (Reid 1993, II:281).

It was during the late sixteenth and first half of the seventeenth centuries that a joint Makassarese-speaking kingdom rose to dominance on Sulawesi's southwest peninsula, as agrarian Gowa combined with its sea-oriented neighbour Talloq to create the combined state of Gowa-Talloq. Both the port-city and the state were known to foreign traders as 'Makassar'. The kingdoms converted to Islam in 1605, and spread the faith along with their own hegemony over much of Eastern Indonesia. Inevitably, Makassar came into conflict with the VOC which was intent on controlling the nearby Spice Islands of Maluku (the Moluccas). Makassar was finally conquered in 1667–1669 by an alliance of the VOC and local forces, which included Arung Palakka, a prince from Gowa's main Sulawesi rival, Bugis-speaking Bone.[11] The Gowa-Talloq Sultanates had indeed been crippled, but how devastating was this defeat to the wider economic and political systems within which Makassar had played such a crucial role?

Lieberman's suggestion (1993:478), that 'the heavy emphasis on maritime influences to explain local change [in Southeast Asia] tends to be reductionist and exaggerated', could usefully be applied to Makassar. As William Cummings (2002:22–23) writes:

> Perhaps because of the important place Makassar occupied in archipelagic trading networks, the concomitant Western tradition. [sees] Makassar primarily as a port... The most sophisticated and recent such treatment is Anthony Reid's *The Age of Commerce,* in which Makassar

11 The Bugis were the other, larger, ethnic group on the southern peninsula (L. Andaya 1981).

> plays a key role… Not wrong, but rather one sided, this work exemplified how Western scholars in particular have understood and described Makassarese history.

Unlike the archetypical port-city of Melaka on the eponymous Straits, Makassar was not particularly well located for long-distance shipping. However, it did have a rice surplus under royal control,[12] and rice was Makassar's only significant local export. In fact, Makassar's image as a 'cosmopolitan port city' should be supplemented by another, at least as important: of Makassar as the centre of an agrarian tributary system.[13] In such polities products were extracted from a subject peasantry through chains of chiefs, each of whom took his cut. The rest went to the courts. Country rice fed the nobles, fighting men and slaves clustered in the city's compounds, and also entered the market, especially when the rulers and nobility sold or bartered their tributary surplus. Foodstuffs and manpower from the interior sustained fleets and armies; we should remember that for most of the 'Age of Commerce' Gowa-Talloq was on a war-footing (Cummings 2002:31–32).

Although sea-farers from around Makassar had been visiting Melaka in the early sixteenth century, Makassar itself was not then an international port-of-call. Javanese and Malays came to trade, but Melaka was the regional hub. Then, in 1511, Melaka fell to the Portuguese, and again to the Dutch in 1641. After each defeat well-connected refugees fled to other ports, including Makassar, bringing commercial expertise, access to capital and networks. Moreover, by the 1620s the Dutch were restricting free trade in the Maluku Spice Islands, and Makassar became an attractive alternative source of cloves, nutmeg and mace. These drew more Portuguese, Chinese, Englishmen, Danes and Indians to the port, where they exchanged goods from Macao, Manila and Cambodia, and textiles from India, for regional products (Noorduyn 1983; Sutherland 2004). It is this early seventeenth century traffic which exemplifies the idea of Makassar as a cosmopolitan port.

In fact, most of Makassar's trade was regional. A couple of hundred ships sailed the Sulawesi, Banda and Java seas, while an estimated twelve went to nearby Kalimantan (Borneo). About a dozen undertook longer journeys to the north (Macau, Manila and other Philippine ports) and a similar number went

12 The consensus is that after c. 1400 rice production and population growth in South Sulawesi intensified, as agriculture came to offer a stronger base for state formation than trade (Macknight 1983; Druce 2009; Bulbeck and Caldwell 2008).

13 Wolf (1982:82) emphasizes that such tributary 'modes of production' could not be understood in isolation, but only in the context of power constellations.

west (to Siam, Cambodia, Melaka and Aceh). These twenty to thirty vessels were primarily owned and sailed by immigrants, but a few high nobles also participated (Knaap and Sutherland 2004:19). The court used its political clout and links to a handful of allied foreign merchants to dominate access to high-value imports, such as American silver from Manila, and textiles from the Coromandel Coast (Sutherland 2004). But the Makassarese themselves were not traders; Malay and Bandanese refugees, and the Bugis Wajorese from Sulawesi's east coast, were the major shippers. Creole Portuguese also used Makassar as a base in their voyages between Macau and the sandalwood island of Timor.

After the conquest foreign merchants were expelled, and Makassar's commerce became provincial. Long-distance trade to the Straits of Melaka and the importing of luxury goods and war supplies was restricted by and to the Dutch East India Company. However, the VOC's ally, Arung Palakka, became ruler of Bone and developed his own commercial networks. After a difficult few decades, in the early eighteenth-century, the growth of trade with China enabled Makassar to re-create its pivotal role, albeit in north–south, rather than east–west traffic. Nonetheless, trade volume was comparable to that of the earlier period. Some routes were strengthened, though the range of voyaging was much reduced (Knaap and Sutherland 2004:21, 170). 'Smuggling' to the west grew sharply from the 1720s, as Bugis and Chinese networks expanded (L. Andaya 1991). The Dutch Company's defeat of Makassar had been a very severe blow to local trade, but within the wider, regional commercial system new links were re-established and old connections revived.

The VOC cut the political heart out of Makassar, but the reconstituted settlement had many similarities with the previous town. Gowan Makassar had undoubtedly been a pre-modern city, with a relatively dense population, a core of fortified settlements and diverse neighbourhoods. The cultural nucleus of courts and mosques provided a model that was influential throughout the region (Cummings 2002). After 1669 Makassar's wider exemplary role was assumed by Bone. But, aside from this brutal fact, was the Company's takeover as devastatingly absolute as Reid suggests? Was Gowan Makassar really so large and urban, and the VOC town so puny? Reid credits old Makassar with a population of over 100,000,[14] but I would suggest the actual number was perhaps half of this. There are two main reasons for this difference of opinion: the first concerns the location of city boundaries, the second the actual calculations as to the number of inhabitants.

---

14 Figures for populations and city area evolve through a series of publications, see Reid 1980, 1993, II, 2000b.

The core area of the Makassar kingdom stretched over the present coastal plain of the Jeneberang River and the broad valleys of the main tributaries of the Talloq River.[15] The archaeologist David Bulbeck has estimated the mid-seventeenth century population of this at c. 300,000, divided between three zones.[16] Roughly equal numbers are attributed to the city, the belt of neighbouring villages, and outlying districts. The first area was centred on the Jeneberang estuary and stretched c. 3.6 km inland; this Makassar city has been described as 'a collection of hundreds of villages located around the forts'.[17] Bulbeck, whose own research focused on the outer zones, adopts Reid's lower estimate of c. 80,000 for this urban centre.[18] The territory of the contiguous villages extended between 3.6 and 8 km from the coast, and supported c. 90,000 people with an estimated density of 12.25 people per hectare, higher than that of the coast itself (10.47). This suggests, notes Bulbeck, a 'suburban' rather than a rural setting.[19] This flourishing hinterland was divided between the two main kingdoms, Gowa and Talloq, each with c. 44,000 inhabitants. The third and outer region of some 98 square miles, extending c. 7–20 km inland, supported c. 110,000 people.[20]

In this context Jan Wisseman Christie's description of Java's 'states without cities' is very relevant. She writes:

> The underlying pattern of the growth of population in central and east Java seems to have been, during the pre-colonial period, that of short term fluctuations in very mobile urban populations balanced against long-term (though not necessarily steady) growth in density of rural populations in core regions of states. Not only did villages grow at the expense of larger enclaves, but the data suggests that as villages grew they tended to break down into two or more nucleated hamlets rather than acquire

15 Bulbeck (1994: Figures 3–6); Reid (1993, II:82). In 1987 Reid viewed the city as a 10 km stretch of coast from Panakukang right up to Ujung Tanah, like the modern city, but in 1993 he defined it as 6 square kilometers lying well south of Fort Ujung Pandang.

16 Bulbeck (1994). This revises conclusions in his important dissertation (Bulbeck 1992).

17 Mangemba (1972:7); Reid (1993, II:82).

18 He now feels that 50,000 might be closer to the mark; personal communication, October 2012.

19 In the 1980s the pioneering scholar of Southeast Asian urban studies, Terry McGee (1991), described rural–urban agglomerations (*desakota,* 'villagecity') 'that posed particular challenges to urban theory'.

20 Bulbeck (1994:4–6), Tables 2, 3; Figure 2.

> the characteristics of small towns. Most of the towns and cities of modern Java owe their size, and many their very existence, to Dutch intervention.[21]

The relative importance of the 'suburban' zone complicates the distinction between urban and rural. Reid (1993, II:85–88) has always recognized that early Southeast Asian cities were 'green and sparsely settled' agglomerations of 'fortified urban compounds', with 'little concept of a specific urban space'. Combined with Bulbeck's findings, this suggests that a spectrum is a more appropriate metaphor for such settlement clusters than a simple city-country dichotomy. Similar gradations existed in economic roles, particularly within the coastal or inland suburban zones. Some traders, for example, would have been full-time shippers, whose expeditions could last anything from a few months to several years. But a majority of sailors would also have been fishermen or farmers, who probably only took to the boats for shorter seasonal voyages once a year. Many women would have had similar manifold roles, participating in cultivation, market trade and household weaving. Moreover, most nobles would have had their power base in the countryside, and their primary residences may well have been rural.

If the urban core of Makassar, the 'city', is restricted to those whose livelihood did not depend completely on agriculture an estimate of about 50,000 seems generous. This would include political centres,[22] markets, maritime- and land-trading communities, as well as foreigners.[23] This figure is based on rather arbitrary estimates: c. 18,000 living in compounds led by men primarily concerned with politics; c. 20,000 in fishing, shipping and maritime commercial households; artisans, market people and land-trading, c. 8,000; c. 4,000 foreigners (excluding Malays, who could be considered local). Extensive areas of garden and orchards lay within the town, but purely agrarian villages

21 Christie (1991:24–25). For further discussion on cities in general, see Sutherland 2005.

22 Bulbeck (1994:6). The two main palaces covered respectively 84 and 44 hectares, but were not necessarily densely populated; 'Somba Opu, the most important, contained only the dwellings of the royal family and their retainers'.

23 Rough calculations are based on i.a. the numbers for eighteenth-century ships crews, as the volume of trade was similar to that before 1667 (Knaap and Sutherland 2004:62, 83). Archive references include early eighteenth-century figures for the noble household of Arung Teko (Nationaal Archief, The Hague, VOC collection (hereafter NA, VOC, volume and folio numbers) 1663/174–186) and the population of kampung Melayu, early nineteenth century: Arsip Nasional Republik Indonesia, Makassar Residency archive (hereafter ANRI, Makassar Archive, bundle and document numbers) Mak. 289/2.

in the hinterland and interior would be excluded. If they are not, the characterization of urban becomes very vague, and boundaries impossible to establish.

As in Christie's Java, Makassar's population would have fluctuated considerably. At times the court summoned subject lords to the city, with their very large retinues, while war concentrated or dispersed populations. Fighting, famine or disease could lead refugees to seek shelter in the centre, or safety in rural areas. Reid's population estimate of over 100,000 in 1636 is based on the assertion that 60,000 were then said to have died in the plague, a normal enough ratio. However, this loss of life is not reflected in his subsequent still higher number for 1660, which is based on a particularly unreliable source.[24] If we disregard this 1660 figure, then it is possible that the settled population of Makassar (ignoring temporary mobilizations) was already halved before the Dutch conquest. Nonetheless, it is clear that after 1669 the Makassarese population was decimated by famine, Bugis mass exports of newly created slaves, and large scale voluntary emigration (L. Andaya 1981:208–227).

At first glance, detailed Company censuses for post-conquest Makassar seem to confirm Reid's suggestion of a very dramatic decrease in the number of inhabitants (to 'no more than a quarter' of the pre-1660s level).[25] After all, the early VOC establishment rarely exceeded 600, mostly soldiers, which was a fraction of the previous regimes concentrations of court retainers and warriors. Even when the large slave contingents, and Chinese, Malays and other settlers in the Company town were counted, the total would have been much, much lower than that of the earlier settlements. In 1676 the Dutch-governed population was only 1,382. In 1740 there were 2,128 members of European households (mostly slaves, Asian and mestizo wives and children), as opposed to 2,708 Asians in the Dutch governed *kampung,* or semi-rural villages. By 1790 the figures were 2,706 versus 4,934,[26] still perhaps less than 20% of the previous population. However, comparisons between the pre- and post-conquest situation must take into account the implications of Makassar's dismemberment in the late 1660s.

After 1669 both the peninsula and Makassar's core territory remained, as they had always been, a mosaic of different nuclei and jurisdictions. The VOC ruled the enclave of Makassar, nearby rice-producing provinces and the seas, while the land was (increasingly) dominated by the Bugis kings of Bone.

24 Reid (1993), II:72. Nicholas Gervaise based his account on the memories of two young Makassarese exiles in Paris. He was, according to Lach, 'rather gullible' (Lach and Van Kley 1993:1448).

25 VOC censuses were based on detailed lists compiled by the heads of urban quarters.

26 NA, VOC 3939/123.

Company censuses only covered their own subjects, clustered in the castle and adjoining villages. Followers of independent rulers were excluded, even when they lived in the same *kampung.* Arung Palakka, Dutch ally and later Bone king, had been granted land, villages, rice-fields and salt-pans within the area of old Makassar, and he soon ruled most of northern Makassar from his palace of Bontoala. His successors also lived much of the year in the city, at Bontoala and Rompegading. When the Bone king was expelled from Makassar in 1814 about 10,000 of his people were said to have left.[27] Those who remained, or returned, became Dutch subjects. Consequently the counted population more than doubled between 1790 (7,640) and 1816 (16,677).[28] In 1828 it reached 19,007, including the sparsely settled off-shore islands.[29] This would suggest that in the later eighteenth century Makassar's independent Bugis may have been as numerous as Dutch subjects. Moreover, although the Gowa and Talloq courts had lost the port zone, their rice-rich hinterlands and previously conquered polities, they continued to control some shoreline and farming settlements. These rump territories, geographically separated by VOC lands, were also excluded from Dutch censuses (Stavorinus 1797:198).

This political fragmentation ensured that Company 'soul-counts' in post-conquest Makassar' core area only included part, probably less than half, of the territory's actual inhabitants. So while the total population would still have been smaller than in the comparable Gowa-Talloq area, it was not, I believe, nearly, as dramatically reduced as has been supposed. Despite formal separation, Makassarese, Bugis and Dutch settlements clustered together, and often intermingled, over much the same territory as the old urban core. Cycles of de- and repopulation, like those of the 1660s, were always common in South Sulawesi.[30] In any case, by the late eighteenth century the Dutch ruled zone was (once again?) densely occupied, with most rice fields controlled by Makassarese or Company subjects (including Bugis), leaving immigrants with little access to land.[31]

---

27 ANRI, Makassar Archive, Mak. 353/2.

28 Olivier (1834–37:199–200).

29 ANRI, Makassar Archive, Mak. 3/1.

30 This was very clear in the case of the rice-rich Northern Districts, see also Raffles (1830: Appendix F). He estimates the southern peninsula's population (c. 1815) at half a million, but concludes it must have been much higher given the evidence of abandoned settlements.

31 Beth memorandum (ANRI, Makassar Archive, Mak. 289/2). In the 1770s the Makassar area consisted of the core of old and new Gowa and the settlement of Makassar itself, which extended from 'Sambung Jawa [a kampung inland from Tanjung Bunga, just north of the Jeneberang] out to the so-called *kraal,* then to the north along the salt-pools behind

The extent to which Gowa-Talloq's Makassar could be uniformly described as 'urban' remains very uncertain. Classic definitions of cities emphasize the difference between town and country and the development of specific institutions. But Makassar's boundaries were blurred, and in all probability most aristocratic compounds and shoreline villages were not so very different from those in the interior or further along the coast. There is clear evidence, however, that at court there were some sophisticated men actively absorbing and adapting unfamiliar ideas, ranging from mathematics to Portuguese military techniques (Reid 2000a). The rulers themselves were open to new religions; they considered Christianity before converting to Islam in 1605.[32] This was then imposed on subjects and conquered states alike. However, it is unclear how far such open attitudes, and the new religion, typified the town, as opposed to just (part of) the nobility. Indeed, it is quite possible that ties between like-minded people in separated settlements could be closer than local cross-class or intercommunal connections. This would apply to immigrant communities, as well as nobilities intermarrying across borders.[33] In such cases contiguity was trumped by systemic exchange, often buttressed by kinship and religion.[34]

Within truly urban societies synergy between different groups creates an integrated culture distinct from that of the peasantry. Indeed, Reid (1988, I:89), commenting on Makassar's Islamization, suggests: 'An unusually rapid transition occurred...perhaps because at the time...it was already a cosmopolitan city socially "overdue" for change in a bourgeois direction'. This sweeping statement conflates the courts' acceptance of Islam with a city-wide religious shift, linking complex concepts – 'cosmopolitan', 'bourgeois' and 'conversion' – in a simple causal chain.

Even if bourgeois is used in the loosest sense, to designate a merchant group, we should remember that in Makassar such men were individual clients of powerful nobles, not representatives of an established class. Oliver Wolters (1999:42) commented that 'Southeast Asian cities were not Venices or Genoas. They were royal centres, with trading ports under their shadow'. Jeya Kathirithamby-Wells (1993:136–137), writing of Southeast Asia as a whole,

---

Bontoala, up to the river of Patinggaloang, not far from Ujung Tanah' (Stavorinus 1797:198). 'New Gowa' was the emerging royal settlement at Jongaya, just across the river to the north of Sombaopu.

32 Pelras (1985). Makassar's conversion was somewhat later than many other regional kingdoms; Makassar's rival Ternate became Muslim in the last quarter of the sixteenth century.

33 On the close personal mid-seventeenth-century friendship between the highest nobles and the Portuguese merchant Francisco Viera, see Boxer (1967).

34 For a contemporary analogy see Sklair (2001).

reminds us that '[f]ree market forces and security for property and wealth, vital ingredients for commercial initiative and property accumulation, were not to be found within the framework of royal absolutism and monopoly'. Moreover, in early Indonesian towns production was unimportant, and it is for that reason that Boomgaard (1989b:317–344) downgrades Reid's estimates for levels of urbanization even in densely settled Java. Cummings revisionism, with its emphasis on syncretic practice, gives a more convincing account of conversion (Cummings 2001, 2007). Elsewhere Reid (1993, II:328) concludes that 'The most obvious, but also the most profound, changes of the Age of Commerce were in the areas of religion and mentalite'. He adds that the later 1600s were marked not only by 'a retreat from the international market but also a distrust of external ideas'. We have so little information on the internal intellectual history of late seventeenth and eighteenth-century Makassar that we cannot judge, but there is no real reason to assume that the indigenous elites, let alone the expanding Sino-Malay settlements, were more insular. Reid's own caveats deserve more emphasis.

In fact, it seems that the fall of Gowa-Talloq was a 'political crisis at the level of states'[35] rather than a systemic collapse. If Makassar's history is considered in isolation then its defeat in 1669 was catastrophic, and Gowa itself did indeed become an embittered and commercially insignificant kingdom. But if, on the other hand, Makassar is seen as a node in one or more systems, then systemic shifts are much less dramatic. There was no definitive decline in indigenous power or in regional commerce. On the contrary, Gowa's defeat marked the beginning of Bugis Bone's unparalleled rise. Wajo and Bone took over much of the regional trade, following the logic of supply and demand. Moreover, illicit connections with the Straits, Chinese and Western private traders were encouraged by the many new coastal settlements established by Makassarese and Buginese who fled Makassar after the conquest. From the 1720s Bugis-influenced Riau at the foot of the Straits was the most prominent link, but smaller Bugis groups throughout the archipelagos played similar, if more modest roles. After 1746 annual junk visits from Amoy (Xiamen) intensified Makassar's role as a pivotal intermediary between southern China's ports and the archipelagos' suppliers of marine and forest commodities. In the course of the eighteenth century the city recovered. For South Sulawesi, and Makassar, the 'seventeenth-century crisis' was a period of adjustment, as had happened before, and would also happen again. This long-term perspective, emphasizing Makassar's role in political and commercial systems, shows that Makassar's decline in the 1600s was not so exceptional, nor was it definitive.

35 This phrase is used by Subrahmanyam (1997). See also Sutherland (2005).

Looking Back: Reconstructions of Makassar

Nowadays Makassar's government regards the late sixteenth and early seventeenth centuries as exemplifying the city's essential identity. The official city web-site[36] notes that:

> [T]here was an intensification of local, regional and international trade, in the political sector and in the physical construction of the city. [...] In the space of less than a century Makassar became one of the most prominent trading cities in the world, inhabited by more than 100,000 people. with a cosmopolitan and multicultural character. Until the middle of the seventeenth century Makassar was able to consolidate its power over a large part of Eastern Indonesia. [.] Internationally, as part of the World of Islam, the Sultan of Makassar maintained close trading and diplomatic ties with the kingdoms of Banteng [Bantam] and Aceh in West Indonesia, with Golconda in India and the Ottoman Empire in the Middle East.

Here Reid's account is accepted and amplified. Makassar is removed from its context and presented as being at least the equal of contemporaneous European cities. In this seventeenth-century Golden Age Makassar epitomizes South Sulawesi's maritime and religious virtues, the glory before the fall. Official narratives of subsequent eras are much vaguer. A generalized emphasis on sea-faring traditions and Islam is followed by isolated accounts of anti-colonialism. This approach neutralizes potentially sensitive questions, such as warfare between states, the 'feudal' and exploitative nature of local polities, the strong Chinese economic presence, conflicts over religious practice, and wide-spread factional fighting between 1945 and 1965. This official history also confirms Makassar's natural right to be the commercial centre of East Indonesia. All in all, it serves its purpose very well.

From the late 17th to the mid 20th centuries Makassar, a Dutch island in a sea of independent states, lacked the anti-colonial credentials to feature in these political narratives. The city was largely irrelevant to early Islamic reformism, which was concentrated in the port-town of Pare Pare and the kingdoms of Wajo and Bone.[37] Makassar re-enters nationalist narratives with the founding of the city's first modern political and religious movements in the early twentieth century (Poelinggomang 2004; Harvey 1974). Even so, the colonial

36 http://bahasa.makassarkota.go.id/index.php/sejarah-kota-makassar (accessed 23-3-2014).

37 Mattulada (1976); Gibson (2000, 2005); Van Bruinessen (1991); Pelras (1994); Hamonic (1991); Halim (2004).

enclave was insulated from regional politics until 1942, when the Japanese abolished the city's special status. Bone, not the town bureaucrats, became their key political interlocutor. Dutch Makassar was again relatively isolated from the countryside during the turbulent periods of revolution (1945–1949) and rebellion (1945–1965). After the post-1949 Dutch withdrawal city politics changed. The old patterns of bureaucratic and communally-based negotiation were replaced by the rough and tumble of peninsular politics. These were still shaped by pre-colonial patterns, which is hardly surprising as full colonial rule had only been established after 1906. Recent anthropological accounts focus on competition and status,[38] while political scientists unravel personalized contemporary power-plays.[39]

In the rather sparse economic history of South Sulawesi, Makassar's marginalization is reversed, it's commercial centrality emphasized. The more emotional accounts identify trade, and particularly the local sailing fleet, with vague but heroic maritime traditions.[40] Regional context is often lacking. Developments inland or in regional seas are ignored in favour of international linkages, despite the fact that Makassar is presented as 'the Gateway to Eastern Indonesia' (Poelinggomang 2002). The corollary, that it was always a second-tier entrepot of contingent prominence, is less commonly highlighted. In fact, for over 75 years after 1669, Makassar's traffic was focused on Batavia, until its own link to Amoy developed in the second half of the eighteenth century. The resulting China-based prosperity collapsed in the face of Singapore's success after 1819. Although trade between Sulawesi as a whole and the British free port grew, Makassar stagnated (Wong Lin Ken 1960). This decline was reversed for a few decades following partial liberalization of Makassar's trade in 1847. Yet it continued to be outclassed by Singapore, Penang and Surabaya, and after 1873 it was further undermined as newly opened regional ports claimed greater shares of long distance traffic. Only late nineteenth-century political intervention revived Makassar's fortunes, as a hub for the KPM shipping line from 1891. Then the creation of a regional customs union in 1906 confirmed Makassar's pre-eminence, as the colonial military eliminated local kingdoms' political and commercial independence (Poelinggomang 2002; A Campo 2002). The following few decades of relative peace and prosperity ended in the political crises and smuggling economy of the mid-twentieth century (1940–1965) (Harvey 1977). Subsequently, despite heavy investment since the 1970s, Makassar port has remained under-utilized, while Singapore and emerging regional ports

38 Chabot (1996). More controversially: Errington (1989); Caldwell (1991).

39 Magenda (1989); Buehler (2007, 2009); Tomsa (2008).

40 In fact the majority of trader-sailors have always been Buginese, rather than Makassarese.

go from strength to strength.[41] Governments have often tried to channel traffic to Makassar, but such efforts usually proved ineffective.[42] Makassar's exports are now primarily local cash crops, although it remains central to the illegal trade in exotic regional marine and forest commodities.[43]

Conclusion

For centuries Makassar has been a participant in multiple, highly competitive political and commercial systems, only parts of which are documented. While successive regimes strove to control commerce, geography and entrepreneurial ambition ensured that 'smuggling' was constant, if largely invisible. Illicit trade circuits went undocumented under the VOC, and even in the mid-nineteenth century as much as half the port's traffic was not registered in the main statistical series (Sutherland 2013). Similarly, as we have seen, Dutch eighteenth-century sources missed a large proportion of Makassar's population, while pragmatic negotiations achieved political compromises that were better left unreported.[44] All parties – Dutch officials, Bugis rulers, local traders, Chinese and Malays – were tied into wider networks, and represented interests that were shaped by distant as well as local priorities. The history of Makassar – town, state and region – can only be understood if these diverse contexts are reconstructed as clearly as possible. If we accept the partial picture that survives in the documents as the full story, without testing it against an explicit attempt to model its environment, the result will be woefully curtailed. More important, premature conclusions will be drawn,[45] and fundamentally important questions will be left unasked.

---

41 http://siteresources.worldbank.org/INTINDONESIA/Resources/Fin-Priv-sector/Makassar_Minutes_Final.pdf (accessed 23-3-2014).

42 Sutherland (2004); Dick (2002); Reid (2010).

43 For example, on timber, see 'Rogue Traders: The Murky Business of Merbau Timber Smuggling in Indonesia' (2010), http://www.eia-global.org/PDF/Report-RogueTraders-July10.pdf (accessed 23-3-2014); B. Glaser and M. Glaser (2010).

44 The practice of allowing Bone's representatives to work alongside Dutch customs and tithe collections was later, in the nineteenth century, seen as an example of pusillanimous surrender to Bugis pressure.

45 Huff (2012:4–5), for example, writes without qualification that by 1870 'Southeast Asia had experienced two centuries of de-urbanisation'.

CHAPTER 10

The Expansion of Chinese Inter-Insular and Hinterland Trade in Southeast Asia, c. 1400–1850

Kwee Hui Kian

Contemporary observers had often remarked on the dominant role of the Chinese as trade intermediaries in Southeast Asia during the late nineteenth and early twentieth centuries (Cameron 1865; Furnivall 1939; Robequain 1944). Although the Europeans were the major importers and exporters in this period, they relied primarily on the Chinese middlemen for the retail and distribution of imported manufactured goods and gathering products for export purposes. This phenomenon motivated the Thai government to impose restrictions on the Chinese commercial and other economic activities from the 1920s (Skinner 1957). Similar discriminatory policies were also implemented by the newly-independent regimes in Indonesia, the Philippines and Malaysia during the 1950s and 1960s. Chief among these was the attempt to remove the Chinese trading presence in the district and rural areas of these countries (Golay 1969).

By targeting against the Chinese, there appears to be two underlying assumptions. Either it was thought that this particular ethnic group possessed superior commercial skills or that they had obtained unfair advantages during the colonial era. These implicit assumptions have inspired various research directions in the immediate post-colonial era. They include the scholarship by Clifford Geertz, Alice Dewey and their colleagues to test if indigenous Southeast Asians such as the Javanese possess entrepreneurial skills and why they have lagged behind alien groups such as the Chinese (Geertz 1960; Dewey 1962). They also continue to inform research discussions about how far *guanxi* (networking), *xinyong* (trustworthiness) and other special qualities based on the Confucian creed might explain the superior economic abilities of ethnic Chinese in the late twentieth and twenty-first centuries (Dirlik 1997; Jomo 2003).

Against these assumptions of European sponsorship and Chinese business acumen, this paper argues that the latter's dominant role as trade intermediaries during the colonial period had evolved from developments in the early modern era. It examines how and why the Chinese were becoming prominent players in the inter-insular and hinterland trade of Southeast Asia during the seventeenth and eighteenth centuries. So far little notice has been paid to these commercial sectors compared to the discussion on the maritime trade between East Asia and Southeast Asia, especially the junk shipping between

DOI 10.1163/9789004288058_011

China and the latter region.[1] Although various recent works are paying greater attention to the links between the entrepôt and secondary ports as well as commercial activities in the more rural and remote regions, they tend to focus on specific locales such as southeastern Sumatra, northern Java and southern Sulawesi.[2]

The following section first maps out the Chinese expansion of inter-insular and hinterland trade in Southeast Asia in the period before the mid-nineteenth century. The second and third sections consider the methods of operation of these traders as well as the developments and characteristics of the regional economy during the seventeenth and eighteenth centuries. The argument is that although the Chinese migrants and their mixed-blood and localized descendants did operate in ways beneficial towards small-scale trading, these features were not peculiar to them. Instead what gave them decisive advantage over other groups of Asian and European commercial agents were the specific characteristics in the development of the Southeast Asian economy during the early modern period. This momentum was then carried over into the nineteenth and early twentieth centuries when they were serving as trade intermediaries especially for the European merchant houses.

The Chinese Trade Expansion in Southeast Asia c. 1400–1850

The commercial exchange between China and Southeast Asia dates back to the period before Common Era. The earliest carriers of trade were the Austronesian people followed by the Arabs and Indians at the turn of the tenth century. When they started sailing to Southeast Asia during the twelfth and thirteenth centuries, the Chinese primarily conducted trade at the capital towns of Srivijaya and Champa and also subsequently in Melaka from the early fifteenth century. At these entrepôts, Chinese imports including earthenware, metallic manufactures, textiles and tobacco were sold wholesale to the local rulers and major merchants who were in charge of their retail and distribution (Schafer 1967; Christie 1998; K. Hall 2004).

By the early fifteenth century, some Chinese traders began settling down and opened shops in the entrepôts and other flourishing port-towns such as Gresik, Surabaya and Tuban in north Java as well as Palembang and Siam (Reid 1996). If they could not sell all the goods within their short sojourn at the

1 Blusse (1986, 2011); T'ien (1989); Cushman (1993); Reid (1993, 1994); Reid and Fernando (1996); Ishii (1998).

2 Andaya (1993); Knaap (1996); Knaap and Sutherland (2004); Kwee Hui Kian (2006).

entrepôts, the Chinese junk traders would entrust the remaining commodities to these resident merchants. Besides retailing these goods the Chinese shopkeepers would also buy up products from the locals over several months and sell these to the junk traders and other long-distance merchants when they conducted their annual visits. In areas where commercial opportunities were abundant and the local authorities were welcoming, this settled population became larger. The wealthiest among them also became royal merchants who traded on behalf of the ruler. Some were also appointed as court officials and envoys during tribute missions to China (Reid 1996; Ishii 1998).

Chinese private shipping to Southeast Asia increased exponentially during the late sixteenth century after the Ming emperor lifted the maritime ban in 1567. While 50 junks were granted licenses to trade in the region in the 1560s, the number increased to more than 130 after three decades. Instead of visiting only the entrepôts such as Siam and Melaka, the Chinese junks also started sailing to a range of secondary ports. Since pepper was one of the most coveted Southeast Asian commodities in late imperial China, the latter included especially those located nearest to the pepper-producing regions such as Patani, Brunei, Palembang and Banten (Meilink-Roelofsz 1962; Reid 1993).

From the late sixteenth century resident Chinese traders were not only maintaining shops in the coastal towns but also started itinerant trading. Carrying strings of petty coins and small quantities of imported textiles, earthenware and other trinkets, they peddled into the interior areas to buy products from the agriculturalists. They were especially active in the pepper-cultivation regions in West Java and southeastern Sumatra (Blusse 1986). In Jambi and Palembang, some Chinese traders had also taken up residence in the pepper-growing areas of Kuamang and Tembesi by the early seventeenth century (B. Andaya 1993:54–56, 124–125).

With the arrival of the Europeans in Southeast Asia from the early sixteenth century, the Chinese traders also operated in their enclave port establishments. New to the region, the Iberians and northern Europeans relied on these experienced Asian traders to purchase local commodities for the home market and retail imported items. Within a few years of their establishment the Chinese merchants became dominant players, particularly in Spanish Manila and Dutch Batavia. These two port towns also witnessed a novel development in the history of Chinese migration: While they had previously visited Southeast Asia primarily for commercial reasons, Chinese migrants began arriving at these European settlements not only as traders but also as craftsmen, market gardeners and sugar producers. In Manila the number of Chinese rapidly increased to 20,000 by 1603 and maintained at between 20,000 and 30,000 in the seventeenth century in spite of the Spanish efforts to limit them to a maximum of 6,000 (Wickberg 1965:ch.1). As for Batavia the Chinese population grew rapidly from about 300 in 1619 to more than 3,000 in 1627 (Blusse 1986:80–87).

In spite of these developments the Chinese trade and migration before the mid-seventeenth century was relatively limited compared to that in the late seventeenth and eighteenth centuries. By the latter period the Chinese junk shipping had extended its visits beyond major ports to include the coastal towns of Cochinchina, Mekong delta and Siam in the Indochinese peninsula, Cebu and Sulu in the Philippines as well as Ligor (Nakhon Sithammarat), Sangora (Songkhla), Trengganu, Pahang, Johor, Riau, Siak, Aceh, Brunei, Banjarmasin and Makassar in the Melaka Straits and Indonesian archipelago. At this point, they were not only approaching entrepôts and pepper-producing regions but also the coastal ports located nearest to the areas of rice cultivation, tin mines as well as the source regions of maritime products such as sea cucumber and tortoiseshell (Cooke and Li Tana 2004; Knaap and Sutherland 2004; Blusse 2011).

The spectacular expansion of junk shipping was more than matched by that of inter-insular trade conducted by the resident Chinese as they sailed to practically every part of Southeast Asia by the early eighteenth century. Although the South Sulawesians (including Bugis, Makassarese and Mandarese), Malays, Javanese and other groups of traders still assumed a strong presence in regional shipping, the Chinese migrant traders and their mixed-blood and localized descendants (*peranakan*) were becoming prominent towards the end of the eighteenth century. In Nusa Tenggara for example, the share of Chinese in shipping volume increased from less than 10% in the 1720s to hold an equivalent share of one third with the Malays and South Sulawesians by the 1780s (Knaap and Sutherland 2004:61). As for the trade in the Java Sea, although they were not as numerous as the Javanese shippers, the Chinese had become the most important carriers in terms of shipping volume during the 1770s (Knaap 1996:64).

Besides inter-insular trade, the Chinese and *peranakan* Chinese were also able to extend trade into the hinterland as well as the upstream and more rural and remote areas in most parts of Southeast Asia. The following is a snapshot of where they had gained dominance by the late eighteenth century. These could be divided into five categories: The first of these was in the European trading settlements. It is mentioned previously that the Chinese were the most dominant traders in Spanish Manila and Dutch Batavia at the turn of the seventeenth century. As the Dutch East India Company (Vereenigde Oost-Indische Compagnie, VOC) set up a string of factories and settlements in the Indonesian archipelago during the seventeenth and eighteenth centuries, the Chinese commercial agents followed in their wake. Consisting of between 10 to 20% of the population, the Chinese and *peranakan* Chinese dominated the domestic trade of Ambon and Melaka by the mid-seventeenth century and the trend continued into the later centuries (Knaap 1991; Andaya and Andaya 2001:96–97). Similar developments could be observed in Padang, and the various port towns along the north coast of Java.[3]

In Makassar, they came to preside over the trade of most of the commodities towards the end of the eighteenth century, a phenomenon which Gerrit Knaap and Heather Sutherland (2004) called the 'Sinification' of private commercial sphere in the port-town. By this time, their other competitors in Makassar were only left with niche markets: The *burghers* (mostly retired VOC administrators) played a significant role merely in the slave export to Batavia while the Malay and South Sulawesian traders were only dominant in the trade of locally-produced textiles, iron knives, coconut products and palm sugar (Knaap and Sutherland 2004, ch. 4; see also ch. 9 in this book).

Besides the European enclave settlements, the Chinese traders were also prominent in polities where Southeast Asian rulers granted them special patronage. One of these places was Palembang. An excerpt from Barbara Andaya (1993:124–125) gives a clear idea the commercial scope open to the Chinese in the locality:

> In 1682 there were about thirty Chinese merchants [in the service of the Palembang sultan], four of whom were appointed to supervise pepper deliveries to the [Dutch] Company. ... Below them were four other private merchants who had little to do with pepper and had no influence in the fixing of prices for goods. Twenty-one less prosperous Chinese traders acted as agents for the wealthier merchants in the bazaar and supplied the court with cloth for the interior. At this time there was only one Indian Muslim merchant. Ten years later three of the four men in charge of the ruler's business were still Muslim Chinese...

In Ayutthaya, where many Indian and other foreign merchants used to serve as port-masters (*krom tha*) and foreign ministers (*phrakhlang*), they had lost these positions to the Chinese by the early eighteenth century. By the 1720s, they also served as revenue farmers and controlled the tin trade in Ligor and Junk Ceylon (Phuket) (Pombejra 1993; Ariyasajsiskul 2004).

The third area where the Chinese enjoyed commercial dominance was those where Chinese migrants were doing labouring work. It is seen earlier

3 Dobbin (1983); B. Andaya (1993); Nagtegaal (1996); Ota (2006); Kwee Hui Kian (2006).

that Chinese migrants had been producing sugar in Manila, Batavia and Formosa by the turn of the seventeenth century. From the 1680s and especially during the eighteenth century, they were migrating to Southeast Asia by the thousands. Many of these Chinese migrants were growing pepper (sometimes together with gambier) in Hatien, Chantaburi, Trat, Brunei, Trengganu, Kelantan, Riau, Mentok, Melaka and Penang. Others undertook copper and zinc mining in North Vietnam and Burma, tin mining in Bangka and parts of west coast of Malay Peninsula as well as gold mining in Pulai (Kelantan) and West Borneo. The Chinese miners and agriculturalists' needs for provisions, tools and other supplies were commonly serviced by the Chinese or *peranakan* Chinese traders. The latter also marketed the products of these labourers.[4]

The fourth area of Chinese commercial dominance consisted of places which were producing commodities desired in the Chinese market. These included the above-mentioned pepper-producing upstream areas in Jambi, Palembang and Patani but had also extended to regions producing other commodities such as the tin-yielding regions of Junk Ceylon and Ligor. During the seventeenth century, Chinese merchants were buying tin in Ayutthaya after the vassal lords in Junk Ceylon and Ligor delivered the mineral as tribute to the Siamese king. From the early eighteenth century, the Chinese had established direct links by leasing these tin fields from the Siamese rulers and did smelting there while Malays and Thais dug for the tin (Gerini 1905; Ariyasajsiskul 2004; Reid 2011).

Besides tin and pepper, regions yielding the maritime products of seaweed (*agar-agar*), sea cucumber (*trepang*) and tortoiseshell (*karet*) also witnessed a major extension of Chinese trading links. From the late seventeenth century, the Chinese were getting these sea products from the eastern Indonesian archipelago up to the northern Australian seas, primarily by working with Sulawesian and Malay middlemen. The latter would use the capital advanced by the Chinese to liaise with and buy the products from the Bajo sea people, the specialized turtle-catchers and gatherers of seaweed and sea cucumber. By the late eighteenth century, the Chinese would trade directly with the Bajos, exchanging old iron or lengths of cloth in return for the sea products. In other words there was a gradual development from multi-ethnic collaboration to Chinese-dominated trading network (Sutherland 2000, 2011). Similar transitions could be observed in the Chinese acquisition of birds' nests in Java and sandalwood in Timor (Blussé 1991; Hägerdal 2012).

In fact some regions which used to have a strong presence of Chinese traders seemed to have lost them because of diminishing supplies of commodities

4 Chen Chingho (1977); Heidhues (1992, 2003); Vos (1993); Trocki (1997); Andaya and Andaya (2001); Cooke and Li Tana (2004); Nordin (2007); Reid (2011).

over time. From the late seventeenth century, the Jambi sultan faced increasing poverty due to a decline of the pepper trade. He hence turned to rely on piracy and raiding activities for his income which led to a greater exodus of Chinese traders and further accelerated the commercial decline (B. Andaya 1993:128–129). The fate of the Sulu sultanate was similar. It enjoyed a flourishing trade selling shark's fin, sea cucumber and other maritime products to Chinese merchants during the seventeenth century. By the second half of the eighteenth century, with the depletion of these products in their surrounding seas, the Sulu rulers turned to piracy and slave raiding which drove away most of the trading communities (Warren 2007).

Last but not least, it is also observed that the Chinese became dominant in the rice-bowl regions of Southeast Asia by the eighteenth century. They included central and east Java as well as the Mekong delta, Tonle Sap, Chantaburi and Trat. These spaces drew thousands of Chinese migrants from the 1660s. By assuming authority or leasing whole districts in these regions from the local authorities, the Chinese were able to gain control over rice trade. These traders also readily assimilated with the local population, particularly in the Gulf of Siam, and some of the half-Chinese descendants such as Taksin, Mac Thien Tu and Jayadiningrat also became ruling elite in the local polities (Chen Chingho 1977; Nagtegaal 1996; Cooke and Li Tana 2004).

In these regions, the Chinese and *peranakan* Chinese were able to gain commercial control not only of rice but also other local commodities in the course of the eighteenth century. We know more details about the developments of their trade in central and east Java thanks to the rich VOC archives. During the seventeenth century Chinese merchants were already leasing the rights to birds' nests collection and cotton textiles branding from the *susuhunan* (Javanese ruler) which gave them monopoly over these commodities (Nagtegaal 1996). By the third quarter of the eighteenth century, Chinese traders also came to dominate the trade of the other key products of Java including tobacco, timber and palm sugar, and almost rivaled the Malays in the trade of salt (Knaap and Nagtegaal 1991; Knaap 1996; Kwee Hui Kian 2006). In fact they had effectively penetrated the hinterland economy of Java. For instance, of the 179 passes that the VOC had granted for travel to the interior for tobacco trade in the late 1760s, 90% were undertaken by the Chinese. Some of these Chinese also resided in Kalibeber, then the most important tobacco market in the interior of central Java (Knaap 1996:98, 130–131).

Because of their commercial control of the production region of basic food provisions, the Chinese also reigned supreme in the inter-insular provisioning trade and became the most important traders in the whole of Southeast Asia. At the turn of the nineteenth century only the hinterland trade of a few regions was not dominated by the Chinese. In the Malay Peninsula and Sumatra, trade exchange with the indigenous producers of pepper, tin and gold, as well as with the forest- and sea-dwelling *orang asli* and *orang laut* – gatherers of products such as birds' nests, resins, black corals (*akar bakar*) – was largely serviced by the Acehnese, Malays, Bugis and Minangkabaus (Lee 1995; Andaya and Andaya 2001; Nordin 2007). In Tonkin and Cochinchina, the Vietnamese were the main traders servicing the exchange trade with the Jarai, Rhade, Curu, Roglai, Mnong, Stieng and other uplanders (Li Tana 1998). In these areas, although the Chinese were entering the hinterland trade, these regional traders remained the more dominant commercial intermediaries.

With their trading links extended to most corners of the Southeast Asian region, the Chinese also served as excellent retailers and distributors of imported merchandise. By the seventeenth and eighteenth centuries, besides the Chinese manufactures imported by junk merchants, they were also marketing commodities brought by the Europeans. The latter considered the Chinese so effective in the inter-insular and hinterland trade that they entrusted these intermediary traders with their imported goods such as Indian textiles and opium, either by selling wholesale or granting them the commodities on credit (B. Andaya 1993; Knaap 1996; Knaap and Sutherland 2004).

Methods

In exploring why the Chinese were able to dominate the inter-insular and hinterland commercial realms in Southeast Asia, some historians have noted that these migrants and their descendants enjoyed some advantages because of the specific ways they conducted business. In her discussion about the success of the Chinese in penetrating the upstream regions of Jambi and Palembang, Barbara Andaya (1993) notes their readiness to integrate with the host society. These traders were adept in picking up languages of the natives and often converted to their religions as well. They also commonly married local women who not only helped take care of their shops but also occasionally accompanied the traders in their trips inland to buy up pepper and sell textiles. Andaya (1993:56) has especially remarked on their skills in reaching the producers and consumers directly:

> Using trading methods eminently suited to the local scene, they were essentially small peddlers, being willing to sell by piece rather than lot and offering the buyer a selection of assorted cloth rather than requiring purchase of a twenty-piece pack or *corge*.

Moreover they 'displayed a keen sense of changing tastes' among the interior pepper growers and were content with 'slim profit margins' as well as a 'willingness to carry debts for years before pressing for payment' (B. Andaya 1993:56).

Coupled with the itinerant trade and willingness to accept deferred payment was the Chinese readiness to offer goods and money to the producers and gatherers several months prior to the harvest periods and deliveries. This was their standard practice for securing rice, tobacco and other commodities in Central and East Java during the seventeenth and eighteenth centuries (Nagtegaal 1996; Kwee Hui Kian 2006). The Chinese merchants were also prepared to give credit to the Borneo and Sulawesi traders one year in advance for tortoiseshell and birds' nests in eighteenth-century Pasir and Makassar (Knaap and Sutherland 2004:98–102).

Elsewhere I have also highlighted the importance of the Chinese internal networks and social capital. Chinese merchants based in coastal towns customarily gave low-interest loans or advanced goods to other Chinese traders to retail imported goods and gather local products in the smaller towns as well as more rural and isolated regions. Usually they entrusted the loans and goods to traders from the same home village and district or to those with identical surnames as these forms of symbolic capital allowed for social sanction against cheating behaviour and if necessary for appeals of adjudication by the elders (*gongqin, zuqin*) of these informal organizations. Should the debtor default on the loans or lose the advanced goods it would also be more likely to track down his family members and demand compensation from them. By utilizing these social-cultural mechanisms, a town-based Chinese merchant could extend his commercial links into the hinterland and more remote parts of Southeast Asia (Kwee Hui Kian 2008, 2013).

Although these practices stood them in good stead against other commercial rivals, they were not exclusive to the Chinese. It was observed that other mercantile agents were adopting similar socio-economic strategies as well. The Arab and Indian traders also commonly married local women to aid their commercial ventures in Southeast Asia. In the seventeenth and eighteenth centuries Surat and Coromandel merchants based in Kedah would station their agents in tin-rich Perak to get prior access to the mineral supplies (B. Andaya 1979:67–70). The Bataks, Malays, Acehnese, Minangkabaus, South Sulawesians, Javanese and other regional traders also did itinerant trade into more remote regions and similarly generated advanced credit to secure agricultural products and other commodities (Dobbin 1983; Knaap and Sutherland 2004).

In fact, although they were latecomers in the regional commerce, the Europeans developed these small-scale trading skills over time as well. By the mid-seventeenth century English and Danish vessels would sail along the north coast of Java, putting their traders ashore at each harbour to sell textiles and opium to the local market vendors and fetching them upon return after several weeks (Nagtegaal 1996:117–118). During the eighteenth century the Portuguese from Macao who sailed yearly to Goa, Malabar and Coromandel would also call at tin-producing regions on the west coast of Malay Peninsula. They left money or merchandise there on credit, picking up tin on their return journey (B. Andaya 1979:70).

What one edge the Chinese did appear to possess over other Asian traders was their effective use of small denomination currency. Unless goods exchange was conducted in barter, low denomination currency was needed to service what was essentially small-scale trading. In pre-sixteenth century Southeast Asia, local rulers tended to obtain products through tributary deliveries and taxation while most private trade was done through bartering. As the Chinese began trading in the hinterland of Banten during the sixteenth century they introduced *pici* coins – made of an alloy of base metals such as tin, zinc, lead and/or copper – for exchange with the pepper growers. Since the market value of the metallic content of these coins roughly equated the face value – or what numismatic experts call 'commodity money' – they were acceptable to the local people. By paying the sultan an annual fee, the Chinese *kapitan* (captain or chief of the local Chinese community) based in the port-town of Banten held the exclusive rights to cast *picis* (Blussé 1986).

As the Chinese started developing interior trade in Sumatra, Vietnam, Cambodia, Central and East Java as well as other parts of Southeast Asia, they similarly put into circulation petty coins – either imported from China or other parts of Asia or minted in a nearby port-town.[5] In fact the Chinese were so competent in penetrating local commerce that the Javanese *bupatis* (lords) and the Jambi and Palembang sultans found it more expedient to get rice, pepper and other products through the Chinese trading links than to exact more tribute and taxation. After the Europeans assumed rule over these various port towns they also relied on the Chinese intermediaries whenever the deliveries from local authorities fell short of the demanded amounts (B. Andaya 1993:55–56, Nagtegaal 1996:117–121).

The utilization of petty coins thus gave the Chinese considerable advantage in securing commodities from the producers and gatherers in Southeast Asia. That being said, once these currencies were put into circulation and had gained popularity with the local populace, other traders could use them too. In early modern Southeast Asia, the minting rights were often held by major Chinese merchants based in the port-towns. However all groups of traders – Chinese as

5 Blussé (1986); B. Andaya (1993):80–81, Kwee Hui Kian (2006); Li Tana (2011).

well as non-Chinese – were using petty coins for goods exchange by the second half of the eighteenth century.[6]

In sum, although the Chinese did enjoy some advantages because of their trading practices and utilization of small denomination coins, these methods alone do not explain why this group was able to reign supreme in the Southeast Asian trading world. Instead the reasons should be located in the particular trajectories in the regional economy during the seventeenth and eighteenth centuries. The following section delineates three developments I consider most crucial to explain the rising Chinese commercial dominance in this period.

Context

The first was the increasing European demand for Southeast Asian products and their heavy reliance on trade intermediaries to facilitate these acquisitions. The European mercantile interests operating in the region could generally be divided into the various East India companies as well as private traders. The latter included the English country-traders, Dutch *burghers*, as well as Iberian private traders and their *mestizo* (people of mixed descent and heritage) counterpart. Although many of the Company administrators and private European traders did come to acquire peddling skills and utilize advanced credit system as noted earlier, they usually proved less proficient than the Asian traders at penetrating upstream and hinterland commerce. The European mercantile agents generally abhorred the haggling practices common among the local people especially when they were dealing with fairly small amounts of commodities. Furthermore their Asian competitors proved more capable at enduring the hardship of hawking into the uplands and were also willing to accept a slimmer profit margin. Oftentimes they would sell imported items at lower prices and purchase local products at higher ones than the European traders. The latter hence preferred to trust out their goods to the Asians for retail and marketing (B. Andaya 1993:55–56).

Moreover the East India companies did not muster sufficient resources to handle all their commercial needs in Asia. This was the case even for the VOC which mobilized the biggest number of personnel and ships among the Europeans: Between 1602 and 1796 it sent almost a million Europeans to work in the Asia trade on 4,785 ships, most of them dispatched to Southeast Asia

6 Blussé (1986); Andaya (1993); Knaap and Sutherland (2004); Kwee Hui Kian (2006); Li Tana (2011).

(Gaastra 2003). Until the outbreak of the Fourth Anglo-Dutch War (1780–1784), it dominated the shipping trade in the ports under its authority. Based on the shipping records, the share of the Dutch company in Makassar's traffic had grown from about 25 to almost 45% between the 1720s and 1770s while it claimed more than half of the total shipping volume from the northern Javanese port towns of Semarang and Surabaya by the 1770s (Knaap and Sutherland 2004:68). But the above was limited mainly to the trade between these major ports with the company's headquarters in Batavia. Even including the private shipping trade of *burghers* the amount of VOC manpower and number of vessels were insufficient to transport provisions and other supplies to all its factories, much less to acquire Southeast Asian products from each of the source regions for the European market.

It was, in other words, necessary for the East India companies and other private traders to work with Asian intermediaries whether for provisioning or to acquire merchandise. Within Southeast Asia it would seem most logical to work with the regional traders. However the European companies often waged wars with the indigenous authorities when they were establishing enclave towns such as Melaka, Batavia, Manila and north coast of Java. Native populations were under explicit orders from their disgruntled rulers not to trade with the Europeans or assist in the acquisition of any supplies and commodities, especially in the immediate aftermath of the armed conflicts (Nagtegaal 1996:115–120; Andaya and Andaya 2001:62).

By comparison foreign Asians were less submissive to the orders of the local sovereigns. Hence in the initial period of takeover of Melaka, the Portuguese worked closely with these Middle Eastern, South Asian and Chinese commercial intermediaries. As for the Spanish, they could not trade directly with Asia under the stipulations of the Treaty of Tordesillas they had signed with the Portuguese in 1494. After setting up their base in Manila, the Spanish authorities mainly relied on Chinese middlemen. Since the main commodities the Spanish coveted in Asia were Chinese porcelain and silks, these commercial agents were singularly well-placed to handle the purchases. Champion of the crusading campaigns in Europe the Spanish were also less tolerant towards the Arab and Indian Muslim merchants and preferred to work with the Chinese.

Meanwhile the VOC and EIC – the two other most dominant European companies in early modern Southeast Asia – found it most expedient to work with the Chinese because the Middle Eastern and South Asian traders were also importing Indian textiles and opium for sales and were therefore their commercial rivals. At times they would prohibit Arab and Indian traders from entering the ports they controlled. Otherwise these European authorities would impose high tolls on and/or ban the import of particular types of Indian textiles and other commodities (Laarhoven 1994; Lewis 1995).

Their treatment of the Chinese was vastly different. The Dutch and English commonly 'trusted out' money and cloth to the Chinese who would retail these and gather Southeast Asian commodities on their behalf. The latter included pepper in Sumatra as well as tortoiseshell in the eastern Indonesian archipelago (B. Andaya 1993:55–56; Sutherland 2011). As they were heavily involved in inter-island provisioning especially for the VOC, the Chinese traders also came to dominate the rice and salt trade in central and east Java by the turn of the eighteenth century (Nagtegaal 1996; Kwee Hui Kian 2006).

Besides their antagonism with the local ruling elite, the related attempts by the Dutch to establish monopolies over various Southeast Asian commodities to maximize profits further pre-empted the use of regional traders as their commercial intermediaries. In the course of the seventeenth century, the VOC conquered the Moluccas and Portuguese Melaka and signed treaties with the local rulers in Makassar, Java, Banten and Palembang, Jambi and Padang. The objective of these military and diplomatic ventures was to form a 'closed sea' to protect a Dutch monopoly over fine spices (clove, nutmeg and mace) and other commodities, and to channel key imports of the VOC – Indian textiles and opium – through the company.

In fact once it subjected the north coast of Java and South Sulawesi under its authority, the Dutch administration forbade all regional traders to sail to the Moluccas, especially the Javanese, South Sulawesians and Malays who had been prominent in the spice trade. It also tried to restrict them within the local shipping trade to pre-empt them from establishing links with other European and regional traders in the Sulu Seas, Melaka Straits and west Sumatran coast and from dealing in Indian textiles, opium and other goods over which the VOC had tried to establish exclusive sales rights (Knaap and Sutherland 2004:20–29).

These regional traders did not disappear but re-adapted themselves in the face of these restrictions and challenges posed by the Europeans. Nevertheless while they were forced to shift their trading routes and products in their inter-insular trade, the Chinese could reign most freely in the Southeast Asian seas in the seventeenth and eighteenth centuries. Even though the shipping volume of the regional traders did not decline and even increased at times, it was inevitable that most of the commercial expansion was absorbed by the Chinese traders and their mixed-blood creole descendants. Although the VOC administrators were still wary that the Chinese might engage in spice smuggling, they were seen as the lesser evil compared to the other regional traders. Besides the VOC and Dutch *burghers*, they were the only group of traders allowed to bring provisions to Ambon and Banda. Indeed the VOC regarded the Chinese rice imports to these spice-producing islands as an indispensable service (Knaap 1996:111–112; Knaap and Sutherland 2004:21).

A second development in the early modern Southeast Asian economy was China's rising needs for regional products. Besides pepper and exotic commodities such as tortoiseshell, resins and sandalwood, the Chinese market had started importing rice, timber and other bulk goods from Southeast Asia since the early eighteenth century, especially from the Gulf of Siam, Luzon and other regions closest to South China (Cushman 1993). Moreover there were significant increases in the demand for particular products such as sea cucumber compared to earlier centuries (Sutherland 2011). Hence the late seventeenth and particularly the eighteenth century saw a peak in the Chinese mercantile activities in Southeast Asia or what Anthony Reid (1997) and Leonard Blussé (1999) call the 'Chinese century'.

In this period the Chinese traders also had to utilize more competitive strategies since tin, pepper and other commodities were also increasingly sought after by the European merchants for their home markets and South Asia. Oftentimes the East India companies were able to use their military strength to gain monopoly rights by signing treaties with local rulers or subjugating the territories under their rule. The Chinese did not have such military strength or backing from their home governments however (Wang Gungwu 1991). What they resorted to was to approach local authorities and obtain similar rights to sole purchase or sales of commodities through the institution of tax farming. In other words they acquired monopolies through monetary payment rather than military force. With these leases, the Chinese tax farmers would then dispatch their agents to purchase and sell goods in the interior. In Palembang, Siam and other polities yielding products in demand in China, the Chinese began dominating revenue farming and extending trade into the hinterland during the late seventeenth and eighteenth centuries (B. Andaya 1993; Pombejra 1993).

Because the Chinese traders had the most freedom to move in the Southeast Asian seas, they were able to conduct a kind of relay shipping trade even for goods coveted by the Europeans. For instance, Chinese merchants based in Palembang could transport tin and pepper to Riau, Trengganu, Hatien, Siam and Cochinchina or those based in the latter ports would sail to Palembang to buy these products. Sometimes the trade was conducted on their own account or on behalf of the rulers and aristocrats or with their partial investments. These commodities would in turn be brought by other Chinese merchants to be brought directly to China or to another port closer to the final market. The European East India companies found it difficult to obstruct this trade without jeopardizing diplomatic relationships with the Southeast Asian rulers, especially if the latter had fulfilled their deliveries of the quantity of tin, pepper and other commodities stipulated in the mutual agreements (B. Andaya 1993:188–191, 218–222; Cooke and Li Tana 2004).

By the late seventeenth century the increase in the demand for regional products by the European, South Asian and East Asian markets collectively created immense pressure on the production and gathering process in Southeast Asia. Formerly these commodities had been cultivated, mined or collected using a combination of labour from corvee impositions, slavery, wage system as well as market-oriented entrepreneurship. However these systems had a cap limit depending on the availability of population which was especially sparse in places like the island of Borneo and the Malay Peninsula. The frequency of wars also resulted in depopulation in regions such as Cambodia as well as the southern parts of modern-day Vietnam and Thailand during the late seventeenth and eighteenth centuries.

At the turn of the eighteenth century some Chinese entrepreneurs began to explore new ways to tackle the problem of lack of production labour in various parts of Southeast Asia. They invited migrant workers from South China – sometimes with the pre-approval of local authorities – to produce rice and pepper around the Gulf of Siam (Chen Chingho 1977; Heidhues 1992). The success of these ventures led to the spread of the use of these migrants for cash crop cultivation and mining activities in many parts of Southeast Asia by the mid-eighteenth century. Largely producing cash crops and mining for minerals in demand by the China market, they fanned out widely to regions where the population was limited or unsubmissive and where local authorities were willing to work with them. An estimated total of about a million Chinese migrants were doing labour work in Southeast Asia by the turn of the nineteenth century. The needs of tools, provisions and other supplies of these production forces and exports of their products were usually served by Chinese mercantile agents, who were also often the ones who had leased lands from the rulers for these production activities and acted as liaisons between the labourers and the local ruling authorities (see for example B. Andaya 1993:218–219).

A third development was the growing commercialization in the Southeast Asian production. In some cases they were connected to the burgeoning demands for regional commodities in the European, South Asian and Chinese markets. For instance, the Minangkabaus, Bantenese and Banjarese progressively converted forests and jungles into agricultural lands for pepper growing in Sumatra, west Java and southeastern Borneo because of the rise in international demands in the seventeenth and eighteenth centuries (Dobbin 1983; Knapen 2001; Ota 2006). Meanwhile other developments of market production such as that of Southeast Asian textiles were due to regional economic readjustments. As the Europeans were buying up Indian cloth for the European and other markets, the prices of these commodities became inflated. Unwilling to pay higher prices for the same types of Indian cloth, various Southeast Asian regions including central and eastern Java, Bali, Sumbawa and southern Sulawesi intensified indigenous textile production both for local consumption as well as export to other parts of the archipelago (Knaap and Sutherland 2004; Kwee Hui Kian 2011).

The proliferation of market production generated greater necessities for food imports. By the early eighteenth century the trade in foodstuffs in the rice bowl areas of Southeast Asia – central and east Java as well as Trat, Chantaburi, Mekong delta and Tonle Sap – was in the hands of the Chinese. In the case of Java this was partly because of the sponsorship of the VOC which exerted authority over the north coastal region since the late 1670s (Nagtegaal 1996, Kwee Hui Kian 2006). As for the rice production in the southeastern parts of Thailand, Cambodia and Mekong delta, because its main market was South China from the 1720s and the main exporters were the Chinese, they also began to dominate the domestic rice trade (Cushman 1993; Cooke and Li Tana 2004). Their control of the trade in these rice bowl regions only further entrenched their position as the primary inter-insular provisioning traders.

As they became heavily involved in the intra-regional food trade, the Chinese also developed interests in commodities of commercial value only within Southeast Asia. By the eighteenth century they were dabbling in a very wide range of trade items including slaves, wax, tamarind, areca nuts, *terasi* (fish paste) and cardamom. Towards the end of the century they were able to rival regional groups and became the most dominant traders in products such as the textiles made in Java, Bali and Sumbawa as well as Javanese tobacco (Knaap 1996; Knaap and Sutherland 2004).

Conclusion

In sum, it is undeniable that the Chinese adeptness in gaining local knowledge, effective utilization of peddling skills, advanced credit system and petty currencies as well as strategies of networking through their home region ties, family names and other symbolic capital had helped render them strong players in the Southeast Asian commercial world. However it was ultimately the three inter-related economic developments – the Europeans' almost exclusive reliance on the Chinese intermediaries, China's increasing demand for Southeast Asian products as well as commercialization of regional production – that enabled the Chinese to emerge as the pre-eminent traders during the seventeenth and eighteenth centuries. This was true both in terms of the trade between China and Southeast Asia and between the various islands within the region. They also enjoyed a significant commercial presence in the interior, upstream and more remote areas of most parts of Southeast Asia, gaining dominance in those of Java, West Borneo, South Vietnam, Cambodia and southeastern Thailand.

It would be wrong to imagine that the Chinese had worked as a single ethnic network though. Operating along the lines of specific loyalties based on common home villages and districts as well as surnames, what was effectively at play was a multitude of cross-cutting and competing organizations. Some of these might form temporary alliances at times but more commonly they were competing against one another, occasionally resorting to violent means, as could be seen in the fightings that broke out between rival Chinese traders and miners in late-eighteenth-century West Borneo and nineteenth-century Malay Peninsula (Kwee Hui Kian 2007, 2013).

Nevertheless, taking the commercial operations of the numerous Chinese networks as a whole, they allowed for the effective retail and marketing of goods imported from other parts of the world as well as the purchase and gathering of Southeast Asian products for export purposes. While they were mainly distributing porcelain, iron products and other Chinese wares through these multiple inter-insular and hinterland commercial channels before the 1600s, Chinese traders increasingly undertook the retail of the Indian textiles and opium imported by Europeans during the seventeenth and eighteenth centuries. With the industrial take-off and influx of European manufactured goods from the mid-nineteenth century, European merchant houses also took the cue from the predecessors and tapped onto these established Chinese connections instead of building their own distribution networks from scratch. Expanding on the basis of early modern developments, Chinese merchants and traders were able to gain commercial dominance in the whole of Southeast Asia during the colonial period, a momentum that is carried into the post-colonial era as they also come to service the Japanese, American and other multinational companies.

CHAPTER 11

From Contest State to Patronage Democracy

The Longue Durée *of Clientelism in Indonesia*

Henk Schulte Nordholt

> To seek a protector, or to find satisfaction in being one – these are things common to all ages.
>
> BLOCH 1964, I:147

Indonesia's political landscape is currently inhabited by a large number of civil militias. In Jakarta the biggest militia is the Betawi Brotherhood Forum (FBR) which claims to have a membership of 80.000 people. Competing militias include Forkabi, or Forum Komunikasi Anak Betawi, the Pemuda Pancasila and more than 130 other groups (Wilson and Nugroho 2012; I. Wilson 2012). The existence of militias is not restricted to Jakarta but seems to be a nationwide structural phenomenon. To give a few examples: in Bali Forum Peduli Denpasar (FPD) and Laskar Bali used to dominate the informal security networks of the city of Denpasar, while more recently the Baladika Bali, claiming a member ship of 17.000 people, made its appearance (H. Schulte Nordholt 2007; Wilson and Nugroho 2012). In Kalimantan (Bakker 2009a; 2009b) the Dayak Adat Defence Command is the biggest militia claiming to represent the interests of the Dayak communities, while on the adjacent island of Tarakan various migrant communities have their own militias (C. Wilson 2012). These are just a few examples.

Centered around a charismatic leader and with a membership that is often based on ethnicity, these militias claim to serve the interests of local society and seek to maintain relationships with political leaders. Their semi-military outfits are very similar to the way criminal strongmen (*preman*) operating under the New Order were dressed, and underline the intimidating public appearance of these groups. On the other hand, these militias present themselves as civil society organizations addressing issues like security, social justice, employment, and cultural identity. Many of them are officially registered as mass organizations while they are at the same time involved in illegal activities. Operating in the borderlands of the formal state and rooted in local society they enjoy considerable public support.

In Jakarta FBR claims to represent the interest of the urban poor by resisting evictions of slum dwellers, providing security, and seeking to improve employment opportunities and health facilities for the Betawi population of Jakarta.

DOI 10.1163/9789004288058_012

In Bali the FPD and Baladika Bali present themselves as guardians of traditional Balinese culture against the evils of globalization in general and Javanese thieves in particular, while in Kalimantan the Dayak Adat Defence Command intervenes in land disputes and seeks to obtain lucrative construction contracts. Nearly all militias are involved in election campaigns of political parties, maintain informal relationships with the police, while they receive protection money from café's, restaurants and bus terminals.

The public appearance of many militia leaders reminds one of the 'men of prowess' who once inhabited the fragmented map of small-scale politics of Southeast Asia, and maintained vertical relationships with clients and political power holders (Wolters 1982). Is this just a superficial resemblance or can we trace a continuity stretching from pre-colonial times to post New Order politics in which strong men and clientelism feature prominently? If militias are part of a long historical trajectory going back to the days of the pre-colonial 'contest state', to what extent do their efforts to provide security and justice reveal certain 'state-like qualities' which can also be attributed to their ancestral men of prowess? If we can trace a *longue durée* of clientelist relationships that characterize(d) Indonesian politics, what does that tell us about the present state of citizenship?

In this essay I aim to trace clientelism as a defining feature in Indonesian politics within shifting historical contexts. I will apply the notion of the *longue durée* as a means to overcome the conceptual discontinuities, which are produced by conventional historical periodization.[1] However, in doing so, I do not intend to uncover hidden – and timeless – structures. When Braudel (1980 [1958]) launched the concept of the *longue durée* in his famous essay 'History and the Social Science', he explicitly responded to Claude Levi-Strauss (1968 [1949]) who had suggested that we need to eliminate time in order to discover the elementary structures that organize our way of thinking. Braudel argued that only through the measurement of time (*i.e.* the *durée*) the nature of structures can be identified. He was less clear about the interaction between the various layers of time he had identified, and the extent to which an underlying *longue durée* is affected by cyclical movements and sudden events. By following Peter Boomgaard's approach (2001a) to Southeast Asian history and applying a *longue durée* perspective with regard to key political relationships it is possible to overcome the compartmentalization of historical periodization.

1 I am certainly not the first to do this. Apart from Boomgaard's work in economic and ecological history, and other colleagues in this volume, Day (2002) attempted to trace cultural continuities in processes of state formation in Southeast Asia, while Lieberman (2003, 2009) wrote an impressive history of slow processes of state formation between 800 and 1830 in a wider comparative context.

For, each historical period tends to have certain ascribed characteristics in combination with a particular analytical vocabulary and a specific perspective owned by an authoritative group of experts, which obscures underlying structures and persistent patterns in Indonesian history such as clientelism.

Lords, Followers and the Enigma of the Pre-colonial State

As Anthony Reid (1988) and many others (*cf.* J. Scott 1972) have demonstrated, vertical bonds connecting lords and followers, or patrons and clients, permeated everyday life in pre-colonial Southeast Asia. In the absence of large centralized states, a multitude of lords provided protection and patronage in exchange for loyalty and labour, and surplus and support during wartime. Control over manpower was a vital index of power and status (Reid 1988:129). Because land was abundant but manpower scarce, labour was highly valued and often tied to power holders through debt bondage. Indebtedness was a key mechanism that made clientelism work. Debt relationships were not only imposed by the powerful through punishment, indebtedness, slavery, or the obligation to pay bride wealth, but were also actively sought by the less powerful. Indebtedness was a form of insurance because patrons were obliged to offer protection, while the absence of protection made people extremely vulnerable. Patron-client relationships were personal and intimate and therefore often phrased in kinship terms (Day 2002). Clientelism conceptualized as father-son relations implied a natural hierarchy which could not be denied or contested. Both patrons and clients needed each other's support. Patrons were in constant competition with (near) equals, both among the aristocratic elite as well as at the local level, and clients were supposed to defend their patron's honour.

Michael Adas (1981:218) has characterized the dynamics of patron-client relationship in a context of ongoing competition as a 'contest state':

> Central to this form of political organization is rule by a king of emperor who claims a monopoly of power and authority in a given society but whose effective control is in reality severely restricted by rival power centres among the elite, by weakness in administrative organization and institutional commitment on the part of state officials, by poor communications, and by a low population-to-land ration that places a premium on manpower retention and regulation. These conditions give rise to polities in which there was a constant struggle between ruler and nobility, between factions of the elite at various levels, and between supravillage elite groups and village notables and peasants for the control of

> labor and agricultural production which formed the basis of these predominantly agrarian states. Although the fortunes of the contending parties fluctuated greatly over time, their continuing struggle over revenue control and the inability of any of the parties to dominate the others decisively on a sustained basis suggest the concept of the contest state as a useful way to characterize this form of political organization.

Clients had a relatively strong bargaining power because patrons ultimately depended on their loyalty, which could not institutionally be enforced. Avoidance – not showing up, seeking the protection of another lord, or migration – was therefore a powerful 'weapon of the weak'.

Evidence from Bali confirms this general picture. The Balinese poem *Bagus Diarsa* expressed the special nature of lords and followers in through the metaphor of the herdsman and his cattle:

> subjects are like cattle,
> their shepherd is the Anak Agung [lord, lit.: 'big man'].
> When he feeds them well they are fat,
> their horns are pointed and sharp,
> who dares to molest them?
> It is foolhardy to do so!
> But when the herd is hungry, their flanks fallen in,
> they will travel in unwavering line
> wherever grass is found in abundance.
> If they wander they will be taken;
> What price will be the ransom?
> It is the herdsman who is the fool then;
> he is his neighbours' laughing stock.
>
> H. SCHULTE NORDHOLT 1996:35.

Leaders had to motivate and persuade followers to accompany them to the battlefield. Another source tells how this was done (Worsley 1972:157–159). First, followers were given food and drinks, clothing, weapons and women. Then it was the king's turn to ask to follow him to defeat the enemy and gain wealth and status. Warfare reinforced the system as it urged people to seek protection and challenged strong men to demonstrate their prowess.

Another Balinese poem, the *Kidung Nderet* explains what happened when a lord ignored his obligations towards his followers and started to 'forget' them (H. Schulte Nordholt 1993). When new favourites of the king rose to power and trusted followers lost their well-deserved privileges, internal disputes emerged which weakened the bonds of loyalty on which the king's authority rested. Tensions increased when external enemies took the opportunity to attack and confusion and chaos prevailed. Lesser lords no longer supported the king who faced a humiliating surrender. The lesson of the poem is that when leaders start to forget, the kingdom falls apart which leads to confusion and destruction.

Evidence from South Sulawesi and the Malay world confirms that local politics were informed by similar dynamics. Clientelism formed the core of political dynamics in South Sulawesi. Basic to this system was the awareness that the higher had authority over the lower, an authority that was accepted by the latter and conjoined by the understanding that high and low needed each other to enhance their prestige (Chabot 1950:102). Clientelism covered a wide field of activities, ranging from warfare, hunting, and rituals, to agricultural and domestic work, and trade. A leader's responsibility was to protect and feed his followers, to show generosity and display eloquence and bravery, while clients were expected to support their lords and defend their honour (Pelras 1996:181–186, 2000). In the Malay world everything depended on ties of loyalty that connected the king with various groups of followers. Loyalty and treason were two key words in important texts like the *Sejarah Melayu* and the *Hikayat Hang Tuah* which dealt with kingship and hierarchy (Andaya and Andaya 2001:46–52). A well-known metaphor – similar to one from Bali – explained that subjects were like roots while the ruler was the tree; without roots the tree could not stand upright (Andaya and Andaya 2001:49). The king's position depended on the acceptance of his subjects, and he was expected to consult his ministers before taking a decision, but the biggest sin servants and subjects could commit was treason. It was better to be killed by the king while innocent than to betray him.

In the nineteenth century, when in Europe the bureaucratic nation-state emerged, western observers who tried to understand the nature of 'the state' in Southeast Asia were confronted with an enigma. Envoys sent to Bali in 1840 observed the following:

> Since my first arrival here, I had given myself the task of gaining a picture of the relationships between the rajas [kings]...furthermore about the way of their government, their power etc. The more I learned about it, the more I noticed that I became entangled in a labyrinth of complex family relations and interests. All this became even more complicated by the divergence of information, which was given differently, as one received it from [representatives of various] royal lineages...or anyone else.[...]

> Everything here is chimerical, nothing clear-cut; every day the most blatant contradictions supplant each other.
>
> H. SCHULTE NORDHOLT 1993:291.

'States' consisted of networks of personal relationships, which were hierarchically ordered and often expressed in (extended) kinship terms. The centrality of the king, or *raja*, is illustrated by the Malay term *kerajaan*, which means: the condition of having a king.[2]

There is of course a clear overlap between the metaphor of the family and the Weberian model of patrimonialism. The family metaphor included important notions of origin and offered a model for a hierarchical order in which the core lineage of the king, ideally descending in a direct line from a divine ancestor, formed the centre while affiliated kin groups formed the lower periphery. Marriage and concubinage were key instruments to knit this system together. Origin and hierarchy were expressed in rituals for which large numbers of followers were mobilized and where the worlds of gods, ancestors and men were connected (Geertz 1980). Origin, hierarchy and rituals cannot be separated from personal charisma and the power of the king. Without strong leadership kingship was invisible, which in turn made the kingdom fragile. Seen from the perspective of the clients, royal patrons were not only expected to provide protection against visible enemies, but also against invisible dangers that threatened the continuity of life, like bad harvests, plagues and natural disasters. The notion of potency reflected in this respect both the power of the king and his role to enhance the fertility of his realm. The most important achievement of a successful king was not a destructive display of force but his ability to control violence (H. Schulte Nordholt 1996:143–158).

Continuities and Criminality under Colonial Rule

If we continue this quick, and necessarily superficial, historical overview, the next series of snapshots concerns the colonial period. In the course of the nineteenth century the Dutch gradually build a colonial regime in Java. After they had dismantled the most important royal centres, they spread a thin layer of western institutions over a society that was still dominated by patron-client relationships. Dutch colonial rule was in administrative terms cost-efficient while it produced economically a profitable surplus. The secret of this success was a system of indirect rule which tied the interests of the Javanese regional

2 Milner (1982:114); *cf.* Day (2002:38–39); Nagtegaal (1996:52); Wolters (1982).

elites to the needs of the colonial regime. Both the backing by and the incentives from a strong colonial centre strengthened the position of the regional Javanese elites who were expected to maintain peace and order and to facilitate the economic exploitation of Java by giving access to land and mobilizing manpower. The Javanese elites were clients of the Dutch but at the same time powerful patrons within their own domains. As long as they managed to guarantee peace and colonial prosperity the Dutch did not intervene in local affairs which offered the Javanese elites opportunities to strengthen their regional power. Moreover, since the Dutch made an end to endemic warfare, which had characterized the old contest state, the Javanese elites were no longer threatened by neighbouring competitors.

Under colonial rule clientelism was reproduced but the conditions as well as the precise relationships between patrons and clients had changed. As their position became stronger the Javanese elites distanced themselves from ordinary peasants. An important intermediate role was played by village elites who became the principal clients of the regional elites, while they emerged at the same time within villages as strong local patrons. Both regional and village elites strengthened their position while ordinary clients became more dependent as their bargaining power weakened. A combination of population growth and the ongoing clearance of uncultivated land, which left the population less and less place to hide, resulted in a gradual immobilization of the village population (Boomgaard 1989a; Hüsken 1988, 1994). There were, in other words, fewer opportunities for avoidance protest, which had characterized the old contest state. Instead, protest took the form of violent confrontations which were relatively easily suppressed by powerful colonial forces.

Not all members of the Javanese elite had survived the transition to colonial rule. The Dutch had reduced the number of regional power holders and had virtually abolished an intermediate level of go-between leaders just above the village level. Seen from a formal colonial perspective large districts were now run by a small number of administrators who directly dealt with their village heads. In reality something else happened. The vacuum that had been created by the Dutch was filled by an informal network of strong men *cum* criminals, or *jago*, who operated in a twilight zone. They were used as strong arm and spies by the regional power holders to keep their district under control and they also cooperated with village heads to mobilize labour and to collect taxes. At the same time they were involved in theft, cattle rustling and opium smuggling, while they presented themselves also as local strong men able to protect their village against other *jago* (Schulte Nordholt and Van Till 1999). In their capacity as local power brokers they were sometimes also potential sources of unrest.

In the eyes of the Javanese population the real embodiment of everyday colonialism were not the Dutch who resided in distant places, but the village headmen in tandem with the local *jago*. For them colonialism and criminality were to a large extent synonymous. The Dutch were to a large extent unaware of the kind of criminality they had helped to create. Javanese regional elites, village headmen and *jago* were all descendants of the old contest state. Their relationships had changed since they were incorporated into a colonial regime. Despite these changes clientelism remained a defining feature of rural Java.

Around the turn of the twentieth century large areas outside Java were incorporated into the Netherlands East Indies. Most of these newly conquered areas were brought under indirect rule. As in Java, regional aristocracies played a key role in redefining old patron-client relationships under the new colonial regime. Protected by the Dutch, members of the old aristocracies became more powerful than before in their new roles in which they were expected to give colonial rule a traditional face.

The Dutch colonial regime presented itself increasingly as a well-integrated administrative and a-political bureaucracy, or *beambtenstaat* (Benda 1966), consisting of 100.000 employees (15% of whom were Europeans) who managed a colony inhabited by 60 million people. However, colonial peace and order was not in the first place based on a small and seemingly efficient bureaucracy, but rested primarily on a myriad of informal patron-client relationships which ultimately connected the centres of Dutch rule with Indonesian society.

The Post-colonial Nation-state and the Triumph of Patronage

Revolution and independence created important fault lines because the rise of the new nation-state was accompanied by the emergence of new research centres and new academic paradigms. Initially emphasis was laid on processes of nation-building and national integration, after which the emergence of the strong developmental state and received more attention. These ruptures obscured underlying continuities.

As Herbert Feith (1962:406, 423, 463) indicated in his monumental book on Indonesian politics in the 1950s, attempts to establish a parliamentary democracy were seriously undermined by increasing levels of corruption. Due to rising inflation government salaries went down, which invited bureaucrats to engage in illicit activities, while political parties channelled funds to their own constituencies, which in turn stimulated a process of pillarization of Indonesian politics. This process was reinforced during Guided Democracy (1959–1965) when the national parliament was side-lined by President Sukarno who then headed a patrimonial system characterized by factionalism and intrigues (Crouch 1979). At the top Sukarno tried to maintain a precarious balance between rival groups (communists, Muslims, nationalists), which commanded each a large following in society, while the military formed yet another interest group. The demise of Guided Democracy occurred in 1965 when the military came to power and eliminated the communist party. The end of Guided Democracy and the mass killings of communists marked also the triumph of clientelism.

Carl Lande (1964) was the first social scientist who explicitly applied a patron-client model to Southeast Asian politics to explain the absence of class-based voting the Philippines, while Manning Nash (1965) did a similar analysis in northern Burma. In a general article on the relevance of investigating patronage in Southeast Asia James Scott (1972) indicated that patron-client relationships had survived due to the absence of firm impersonal institutions which guaranteed personal security, while inequalities in society increased because both state institutions and political parties tended to bet on the strong in order to secure support. However, the nature of patron-client relations had changed: relationships tended to become less affective and personal and more instrumental and office based. At the same time chains of patron-client relationship were stretched through political party lines from regional setting towards the capital.

In Indonesia a serious attempt was made to dismantle these chains of patronage and their concomitant social inequality. We follow these efforts in the context of a small provincial town in eastern Indonesia.[3] During the colonial period local rajas had benefitted from the Dutch strategy to bet on the strong in order to establish stable and cheap indirect rule. Due to the Dutch support the gap between the rajas and their client-subjects had widened. After independence the privileged position of many rajas came under attack because the ideology of the new nation-state no longer tolerated old fashioned feudal practices. As a result the power of rajas declined while the new national leaders started to bet on the emerging educated urban middle classes. From this group new local bureaucrats and party leaders emerged who became clients of political patrons at the national level. Employed by the state and with good upward connections with national party bosses they created new and often ethnic based networks of clients at the regional level. They derived their main income from rent seeking through their involvement in state interventions in the local economy.

3 I base this paragraph on new and innovative research by my colleague Gerry van Klinken (2014) on the history of Kupang on the island of Timor.

In this context the Dutch term *kattebelletje* (incorrectly quoted in Feith 1962:479 as '*kaart belletje*') emerged. It refers to a brief informal memo by a patron which gives a client of his access to funding, a job etc. The term slipped into the national language as *katebelece* and resurfaced, for instance, in 1994 in the publicity about the Bapindo corruption case when retired general Sudomo wrote an informal note of support for businessman Eddy Tansil.[4]

In Kupang, the new rent-seeking provincial urban middle class represented a conservative force that allied itself with local businessmen and church leaders. Their position was not uncontested because from the late 1950s onwards the communist party (PKI) gained mass support in its efforts to break the chains of patronage and to spread the idea of equality. As elsewhere in Indonesia the PKI targeted especially the so-called *kabir* (capitalist *cum* bureaucrats). Apart from a growing sense of class consciousness this campaign involved also a new cross-class alliance of progressive middle-class politicians – many of whom were school teachers – with the rural poor.

In terms of actual achievements the results of the PKI campaigns were modest. Its most important contribution was perhaps that it spread the message of modernity by emphasizing progress and equality and by that the promise of a new world that had liberated itself from patrimonial forms of inequality. In the words of one of the party leaders, Njoto: 'We are the men who are modernizing life in the villages; we are the men introducing the twentieth century' (Mortimer 1974:409). Modernity transpired also in the way the PKI was organized. The party was very advanced in organization-building applying universal standards of recruitment, promotion by merit, specificity of roles, a recognized hierarchy of authority and applying uniform roles of control and checking. Observers were struck by the business-like attitude of PKI cadres and the high standards of performing tasks, sowing among its followers seeds of self-consciousness and self-activity. As such the PKI was an important vehicle for merit-based advancement and offered ample opportunities for young talented people (Mortimer 1974:407–410).

It remains of course to be seen to what extent the PKI would have succeeded in making 'a new world' which was no longer dominated by clientelism, but mass murder made and untimely end to these ambitions. Nation-wide more than half a million supporters of the PKI were killed, 6000 of whom lived in the east Indonesian province Nusa Tenggara Timor. It was not by institutional strength, let alone by the rule of law, but through new patronage networks with a new middle class, that the state managed to keep the country together and to reach deep into society (Van Klinken 2014).

4 *Editor*, 24-3-1994.

Strong State and Patronage Networks

Since the middle of the 1970s the rise of the strong centralized state became the dominant theme in many writings on Indonesia. The New Order under President Suharto was described as an authoritarian developmental state, which succeeded to generate sustained economic growth. Richard Robison (1986) was among the first to apply a political economy approach and identified the bureaucratic state as the main motor of capitalist development and the breeding ground of a new capitalist class. In line with a renewed interest in the state (Evans et al., 1985), Benedict Anderson (1983) emphasized in a paradigmatic article also the autonomous role of the state while he pointed at striking similarities between the New Order regime and the late colonial state. Blinded by its economic success – and less critical than the Cornell scholars around Anderson – an increasing number of economic observers – located among others at the Australian National University – were inclined to see the New Order as a rather well-integrated set of institutions that operated apart from society and facilitated ongoing economic growth and de development of a relatively free market (H. Hill 1996).

Harold Crouch (1979, 2010) represented at Monash University a more critical approach towards the seemingly coherent nature of the New Order state, and emphasized the patrimonial nature of the regime.[5] According to him, the bureaucracy was in practice overstaffed, underpaid, largely incompetent, with overlapping tasks and few responsibilities, whereas patronage distribution was the glue that held the system together (Crouch 2010:17). Compared to his predecessor President Suharto had more money to distribute which he derived from oil revenues and foreign aid. This enabled him to control various factions among the elite (military, business, bureaucrats) by granting licenses, access to projects, credit and jobs in exchange for loyalty. A pyramid of patron-client relationships ran down from the presidential palace to the villages, while the entire bureaucracy was permeated with officially encouraged corruption in the context of which government party Golkar operated as a nation-wide patronage network (Crouch 2010:161). There was an informal system of taxation by which bureaucrats supplemented their modest salaries with selling of licenses and levying personalized forms of taxation. In a similar vein the military, whose budget was only partly covered by the state budget, gathered their own funding through business activities and security services.

The extended network of patronage relationships was hidden behind a façade of elaborate bureaucratic procedures and a carefully orchestrated public

5 Dwight King (1982) combined both the bureaucratic and patrimonial nature of the regime.

display of order and unity. National elections were part of this public appearance of order. It was not results that counted but the orderly progress of the campaigns which had to prove that the regime was in control (Pemberton 1986).

At the village level local elites were incorporated in the patronage network of the regime. In exchange for development projects and access to other state resources village leaders were made into clients of the New Order while they were enabled to enhance their role as local patrons.[6] Suharto deliberately emphasized his patrimonial role as 'father of development' by a special development program called Inpres (Presidential Instruction), which suggested a personal involvement with particular projects. Frans Hüsken (1994) observed that towards the end of the 1980s the old village elite in his research area in Java retired and was succeeded by representatives of a new emerging middle class who aspired further upward mobility by gaining access to state resources.

Outside Java, similar processes took place. Pelras (2000:330–334) observed that in South Sulawesi patron-client relationships underwent changes. The old patrons with an aristocrat background were replaced by new educated entrepreneurs who gained wealth in trade and fishing and developed helpful connections with state officials. One of the most successful among these was the Kalla family from Bone. Father and son made a fortune in trade and expanded their business empire trough lucrative government licenses. In 2004 the son, Jusuf Kalla, even became vice-president of Indonesia.

The New Order regime rested ultimately upon violence and intimidation and made use of semi-official militias and informal connections with criminal gangs to harass its opponents. Seen from a long-term perspective one can identify the criminals operating under the umbrella of the New Order as the grandchildren of the rural *jago* in nineteenth-century Java (H. Schulte Nordholt 2002).

Patronage Democracy and Fragmented Citizenship

The monetary crisis that hit Southeast Asia in 1997 paralyzed the patrimonial networks in Indonesia and caused a serious congestion in both the downward flows of material rewards and the possibilities of upward mobility (Sidel 1998). Economic crisis, nation-wide student protests and discord among the elite eventually caused the fall of President Suharto in May 1998.

6 Antlöv (1994); Hüsken (1988, 1994); N. Schulte Nordholt (1982, 1987).

In a very short period of time Indonesia experienced a transition from a centralized authoritarian regime to a decentralized electoral democracy, while the strong state seemingly gave way to the emergence of a strong civil society. However, regime change was not accompanied by the breaking up of clientelism because opposition groups were too weak and divided to press a common reform agenda and although old patronage networks were partly in ruins, many factions of the old regime managed to survive under new circumstances (Crouch 2010; see also Hadiz and Robison 2004). Ultimately the threat of mass unrest from below stimulated timely enough elite solidarity to maintain the basic structures of the old regime. As a result reform measures were allowed as long as these did not fundamentally threaten the vested interests of the elite (Slater 2010:48–50).

At the national and the regional level new forms of patronage emerged, which were headed by unstable alliances of businessmen, bureaucrats and politicians. Due to a far reaching process of decentralization more money flowed to the regional level regional where competing elite factions tried to control these. Contrary to the Philippines where regional dynasties gained a monopolistic position over coercive and local economic resources, regional leaders in Indonesia remained dependent on state resources and vulnerable to electoral changes, and sometimes judicial investigations (H. Schulte Nordholt 2012).

Direct elections created a patronage democracy in which leaders derived their power mainly from the state and maintained relationships with their constituency through clientelistic practices, while their clients identified themselves mainly in local, and often ethnic, communalist terms (Van Klinken 2009:145). At the regional level we can identify the formation of a political class (Vel 2008) consisting of those who dominate regional politics and control access to economic and financial resources of the state which they share with allies, relatives, and clients. A case in point is the rise of the Limpo family in South Sulawesi. Born in an aristocratic family the father started his career in the 1950s in the army, in a time of civil war in South Sulawesi the most powerful institution. Under the New Order he moved into the civilian administration of the province and became active in Golkar while he was also involved in various business ventures. His children followed his footsteps. His son started as a subdistrict administrator and worked his way up in the bureaucratic hierarchy. When he was arrested in 2001 for drugs abuse his career seemed to be over, but the next year he made a remarkable come-back when he was elected vice-governor. In 2007 he was elected governor. Backed by a private security force of thugs he is now heading a wide informal network of business allies and clients.

Decentralization produced on the one hand more state at the regional level, but created also a new polycentric ensemble of power centres. Patronage obstructs in this context a discourse on class differences and does not question inequality. What, then, are the effects of decentralization and patronage democracy on the notion of citizenship?

Ideally liberal citizenship should be seen as the result of a formal relationship between the state and its citizens which is defined in terms of mutual rights and obligations as regulated by the rule of law. Democracy guarantees the accountability of the state while citizenship makes democracy work and vice versa. What happens to citizenship in weakly institutionalized states dominated by clientelism? Seen from an ideal typical definition of liberal citizenship clientelist systems are often defined as 'messy', 'failing' or 'soft'. This approach emphasizes the incapacity of states to maintain the rule of law but fails to explain why so many people profit from these incapacities (Van Klinken and Barker 2009). The answer can be found in the embeddedness of the state in society which forms the breeding ground of many patron-client relationships. Despite important political reforms and a well-functioning electoral democracy, central features of the old patron-client system remain firmly in place in Indonesia because clients prefer strong patrons who deliver more than the weak and unreliable rule of law of an abstract and often distant nation-state. The embeddedness of the state in society reveals also the extent to which middle classes support the image of a coherent state while they seek at the same time access to its resources through personal connections.

To understand the fragmented nature of citizenship in Indonesia it is helpful to follow an alternative approach to the idea of the state from the perspective of the citizen as suggested by Christian Lund (2011). According to Lund, the state defines and guarantees two constituting elements of citizenship: identity and property. Not all government institutions deliver this; hence not all government institutions have state-like qualities. Instead of a fixed set of institutions we should see the state as a relational and often fragmented quality determined by the extent to which certain institutions have the authority to define and guarantee identity and property rights in exchange for which this authority is recognized as legitimate. Citizenship is established by this mutual recognition. Hence, '[w]hen different institutions attempt to govern with state quality, competing processes of consolidation and erosion of institutions of political authority, of formalisation and informalisation of rule, and dynamics of exclusion and inclusion of people unfold' (Lund 2011:75).

In Indonesia many government institutions lack state qualities as a result of which we can observe a fragmentation of citizenship, which produces insecurity. Against this backdrop we now return to the question why militias are so popular. Resembling NGO-like legal entities, militias maintain close ties with power holders and emphasize their role as providers of security which is accompanied by the prominent role of strong men in combination with a military outlook. Instead of threatening the existing order militias claim to support the system while they also claim to represent the interests of their constituencies. Key words dating from the Orde Baru, *antisipasi* and *koordinasi*, are still used and indicate the ever present threat of disorder and insecurity which should be contained in time. Local society is seen as a vulnerable space constantly threatened by outsiders – Madurese migrants are seen as threatening Betawi inhabitants of Jakarta; in Kalimantan Dayak feel threatened by immigrants; 'Islam' threatens Balinese culture (Wilson and Nugroho 2012). In the absence of a strong rule of law embodied in stable government institutions, successful militias have in this context certain state-like qualities by offering security, confirming ethnic (or religious) identities and defending property rights. And they are likely to stay because they are an integral part of the political system. In this respect militias are also part of a *longue durée* of patron-client relationships running from the contest state to current militias, be it through different institutional 'geographies'.

CHAPTER 12

Visual History

A Neglected Resource for the Longue Durée

Jean Gelman Taylor

Felipe Fernández-Armesto enjoins historians to study creative works of the past. Here, he writes, is a precious source of images and sentiments that informed thought and behaviour of peoples long gone (Fernández-Armesto 2002:152). Art and literature, he continues, help historians interpret documents and other material evidence, which are the stuff of historical research. Peter Burke (2001:13) reminds us that artworks, like any other of the historian's sources (which he calls 'traces of the past in the present'), must be studied within their social context. By this he means the cultural, political and material setting, artistic conventions of the day, the circumstances in which an image was commissioned, its intended function, and the physical location in which a work was originally seen. Svetlana Alpers (1983) characterises Dutch painting as 'the art of describing'. Burke links this value for observation of detail with urban culture, and notes that the inventor of the microscope was Dutch.[1] But painters do not see with an 'innocent eye'. Portraits, scenes of small groups, rural and city views are, in Burke's words, 'painted opinion'.[2]

Fernand Braudel's methodology of the long time-span in historical studies of society fosters asking of visual records if there are constants in topics and themes, if there are changes in perspective (Braudel 1980:25–54). In the case of Dutch images of Indonesia and Indonesians, a long time span allows charting of visual experience, impact, adjustment and perspective. It yields a more subtle understanding of what we may already know from written records.

Dutch artists produced a corpus of visual imagery in three centuries of interaction with peoples and places in the Indonesian archipelago. These images formed and reflected sensibilities of artists and audiences in the Netherlands and in communities across 'VOC Asia' and the Netherlands Indies.[3] These latter included the multi-ethnic inhabitants of private households – immigrants,

1 Burke (2001:84). This was Cornelis Drebbel (1572–1633).

2 Burke (2001:19, 122). He adds that 'photographs are no exception to this rule', and this holds true of the so-called 'candid camera' shot as well as the posed or staged photograph.

3 VOC are the Dutch initials for the United East Indies Company (1602–1799). VOC Asia was a string of trading posts in Asian ports and across the Indonesian archipelago, headquartered in Batavia (present-day Jakarta).

locally born, Indigenous – as well as purchasers of artworks, apprentices and artisans in wood- and metal-working, assistants in ateliers and photographic studios, and local artists who came into contact with Dutch artists or their work.

The body of Dutch art in and on Indonesia sprang from an increasingly urban, prosperous society in Holland where, in the seventeenth century, a growing proportion of the population could afford works of art. Art ceased to be constrained by patronage in the Netherlands Golden Age. Alongside small numbers of artists working on commissions for the nobility, guild artists now produced for a mass of anonymous buyers. Art markets and auctions circulated artworks, generated and responded to public demand. In every year of the seventeenth century, 63,000–70,000 pictures were painted for a population numbering approximately 1.85 million (North 2010:90). Inventories of well-to-do villagers and townspeople record ownership of paintings and establish changes in the function of art, popular values and taste. Religious art, which had assisted the private devotions of Roman Catholics, gave way to Protestant preference for morally instructive, mundane scenes. By the 1650s, the Dutch buying public wanted sea- and landscapes, and the well-to-do commissioned portraits of themselves.

This visual culture, with its secular subject matter and love for 'the look of things' (Berger 1974), travelled to Indonesia's islands. Batavia, founded in 1619, already had an art market by 1627. In that year, Gillis Vinant's art collection was auctioned. This Dutch merchant had amassed 28 paintings in all. His collection included nine landscapes and seven Chinese paintings.[4]

Surveying three centuries of production, we find sketches, oil- and watercolours, portraits, sea- and landscapes. Alongside work of the hand, we find mechanically produced images from the earliest days of lithography in the 1840s to photographs and, from around 1912, moving pictures.[5] This creative work catered for a Dutch public avid for images of the Indies. The crowds visiting colonial halls of the great international exhibitions of the late nineteenth and early twentieth centuries attest to the instructive and aesthetic appeal of the image.

4 North (2010:94). Some of Vinant's paintings are identified in the inventory by size and frame rather than subject matter.

5 Many of the paintings and photographs I discuss here are located in the digital image collections of the Royal Netherlands Institute of Southeast Asian and Caribbean Studies (KITLV), the Tropenmuseum (TM) of the Royal Tropical Institute (KIT) and the Rijksmuseum (RM) in Amsterdam. Text citations include archive initials with catalogue number. Websites are: www.kitlv.nl; www.tropenmuseum.nl; and www.rijksmuseum.nl/collectie.

New art forms – wood carvings, illustrated manuscripts, decorated textiles, puppets, ritual objects of gold, silver filigree jewellery, and the like – journeyed from Indonesian manufactories and artists' centres to the glass cabinets of Holland's new museums of the nineteenth century.[6] In this novel cultural milieu, Dutch Everyman and Everywoman could now see for themselves what, until this new age of public culture, had been hidden in the private collections of the royal house and in the country villas of Indies *nouveaux riches* repatriates in Holland. The *longue durée* approach enables us to trace a history of deepening engagement of Dutch people in Indonesian communities over time and place, and a history of Dutch lives there that encompassed relationships of both kin and conqueror. This body of artworks parallels and illustrates research interests of Peter Boomgaard whose scholarship is honoured in this book.[7]

From Ship's Deck to Shore

The first works from Dutch painters were views of Indonesia's bays, fortified settlements and mountains concealing an unknown interior, such as *View of Batavia from the Sea* (RM SK-A-2513) by Hendrick Jacobsz Dubbels (1621–1707). While these paintings are of mighty Dutch vessels, crashing waves and stormy skies, they establish for historians from the beginning the entwined nature of Dutch-Indonesian relations. For the Dutch ships, anchored offshore, are surrounded by smaller Asian vessels ferrying travellers and goods into port.

Going ashore brought the Dutch into relations with sellers of all kinds, introduced women into their contacts, and led to new features of urban landscape in the archipelago. A painting by Abraham Storck (1644–1708), 'Onrust Island off Batavia', (RM SK-A-739), shows dry dock shipyards built to support the VOC's maritime commerce and two windmills.[8] Markets and local characters are recorded in pencil and paint. Some were the work of amateurs, some of draughtsmen and trained artists who drew

6 The world's first colonial museum opened in Haarlem in 1871. Its collections were later transferred to the Royal Tropical Institute, which opened in Amsterdam in 1910. Drieënhuizen (2012) discusses three private collections of artworks assembled in the Indies and later acquired by museums in the Netherlands.

7 Boomgaard's publications cover Indonesia's agriculture and livestock industries, forests, wildlife, landholding, demography, disease and technology transfer. They are supported and informed by visual data from the Netherlands' rich image archives. Boomgaard's *oeuvre* validates insights of historians of art and of the *longue durée*. See, for example, Boomgaard (1996, 2001a, 2003c, 2008a) and Boomgaard and Van Dijk (2001).

8 The flag on the main ship bears the VOC logo and date 1699. Indonesian and Dutch men scrape the hull of one of the ships.

what they saw on site and in their local studio, or who, on returning to Holland, painted from sketches and memory. And some of these early instructive scenes of the Indies were produced by artists who never left Holland.[9]

Four large paintings in the Rijksmuseum collection illustrate Batavia at mid-century. Two are portraits. Much of Aelbert Cuyp's portrait of senior merchant Jacob Mathieusen, his wife and Indonesian servant is given over to Batavia's harbour and fort.[10] The slave-servant holds an enormous *payung* or umbrella over the couple who are about to repatriate. Bay, fort, slave and *payung* evoke VOC Asia, but Cuyp (1620–1691) never left Holland. Yet his painterly imagination captured essentials of its headquarters in his depiction of Dutch and Asian vessels and the Javanese status symbol. Some commentators have scorned Cuyp's depiction of the Indonesian man as fanciful, but the Dutch artist Jacob Jansz. Coeman (c. 1636–1706), who did go to Batavia and spent 43 years there, rendered the male slave in his portrait of the Cnoll family in a style very similar to Cuyp's (RM SK-A-4062). Central figures in this opulent

FIGURE 12.1 *Abraham Storck,* Onrust Island off Batavia
COLLECTION RIJKSMUSEUM, AMSTERDAM (RM SK-A-739).

9 There is no evidence Dubbels ever travelled to Java.

10 RM SK-A-2350. See Zandvliet (2002:181–182) on identification of Mathieusen. Alternative forms of his name in VOC documents are Mahuysen and Martensen.

painting are senior merchant Pieter Cnoll, his part-Japanese wife Cornelia van Nieuwenroode, and two of their daughters. In painting the Cnoll women as Eurasians, Coeman presents an image of the VOC's female elite that we know from birth and marriage records to be accurate (J. Taylor: 2007, 2009).

The other two paintings illustrate Batavia's markets. Albert Eckhout's *A Market Stall in Batavia* presents (RM SK-A-4070), in the manner of a Dutch still life, a rich assemblage of mangoes, bananas, pineapples, rambutan and durian, some cut open to reveal their inner composition. These exotic fruits are named for viewers on a scroll in the painting's right-hand corner. Eckhout (c. 1610–1665) may have derived his notions of Batavia markets from returned travellers, for the scene he paints he never personally saw. Perhaps he viewed tropical fruits grown under glass in Amsterdam's botanical garden, established in 1628. Perhaps he was inspired by his sojourn of 1637–1644 in Brazil as commissioned painter for Johan Maurits, governor of the Dutch colony there. The stall's Chinese seller and Indonesian women customers tell a truth ambiguously.[11] Andries Beeckman (?-1664) painted Batavia's fort and market when in Holland, but based his painting on drawings he had made in Batavia. He painted two versions, each with a different combination of the representative characters who inhabit the central space of his canvas. The Rijksmuseum's version has a pair of robed Chinese men in conversation, a Muslim teacher in white turban and robe, manual labourers in loin cloths and head wrappers, and women in *kain kebaya*, one walking arm-in-arm with a European man.[12]

An early overview of Batavia comes from its Chinese community in the form of a medallion presented to Governor-General Jacques Specx as he prepared to repatriate in 1632 (De Haan 1922–23:G8). It gives an urban planner's view of the walled city, showing the castle's relationship to the sea-lanes, to the young city's neighbourhoods and environs. The obverse is engraved in Latin and Chinese characters. This mapped image of Batavia reminds us that, at this stage, the Dutch were more on Indonesia's seas than on land, more turned outwards than to the villages and princely capitals of the interior.

11 Dutch artists often portrayed servants as cheeky thieves. Cuyp painted an Indonesian stealing a banana from the Chinese fruit seller's stall. Coeman, too, has painted the manservant surreptitiously reaching for an orange from the bowl the female servant offers the Cnolls.

12 *Batavia Castle seen from the West Side of the Kali Besar*, c. 1656, RM SK-A-19. The second version of Beeckman's painting (Tropenmuseum, 118–167) is in the KIT. *Kain kebaya* consists of a long-sleeved blouse and a length of material wrapped around the waist, covering the legs to the ankles.

Dutch Encounters with Asian Arts

The first Asian art the Dutch came to know was a natural circumstance of their positioning within a maritime world: it was the art of China and Japan. The Chinese exported their artworks to VOC markets; the Dutch exported Japan's to Asia and Europe. For all three partners, art was a commodity. Asian and European urban classes wanted to own and display art objects, both novel and traditional. Art in the form of trade goods cultivated individual preferences for a broadening range of creative works. We have already noted the Chinese paintings in Vinant's collection. Probate inventories from Batavia establish that this was to be a characteristic of Dutch collecting in Batavia.[13] Trade figures confirm the taste developed in Europe for Asian ceramics. Over two centuries, the VOC shipped 43 million pieces of porcelain to the Netherlands (De Vries 2008:130).

The Chinese adapted their designs to Dutch forms, such as painted tiles for VOC markets (De Haan 1922–23:D24, D25). In turn, Dutch artists borrowed Chinese motifs and forms. The Amsterdam artist, Cornelis Pronk (1691–1754), for example, sent designs for a series of glazed porcelain plates to his VOC contacts in Batavia. They forwarded his designs to Guangdong and then shipped the finished plates back for sale to connoisseurs of the new *chinoiserie* in Europe around 1736.[14] Oliver Impey and Christiaan Jörg have documented the VOC's export trade in Japanese lacquer ware to other Asian markets and to Holland (Impey and Jörg 2005). VOC officials presented trays, boxes, chests and cabinets as gifts to envoys and potentates everywhere they conducted business, and so introduced a new repertoire into various cultural milieus.[15]

Because VOC dealings with Indonesian statelets were, in the early decades of contact, primarily on the coasts, Indigenous art the Dutch first came to know grew out of Indonesia's Muslim culture. Close to princely residences in port towns the Dutch saw mosques that were distinguished by multiple roofs. VOC-era Dutch were both observers of, and participants in, this Islamic-infused culture. The Dutchman Lucas Cardeel (d. Batavia after 1706), for instance, having become Raden Wiraguna in the employ of Sultan Ageng of Banten (r. 1651–1683)

13 De Loos-Haaxman (1941); North and Ormrod (1998); Scalliet et al., (1999); Zandvliet (2002).

14 Pronk's design has a Chinese woman and maid at its centre, and Chinese men and women alternating with birds in cartouches around the perimeter, RM AK-RBK-15939-A. Pronk's original drawing is catalogued as RM RP-T-1967-18.

15 On Asian luxury goods sent on VOC ships to the Netherlands see Van Campen and Hartkamp-Jonxis (2011).

FIGURE 12.2 *Cornelis Pronk,* Porcelain Dish with Chinese Design
COLLECTION RIJKSMUSEUM, AMSTERDAM (RM AK-RBK-15939).

and a Muslim, designed a minaret for Banten's principal mosque.[16] Minarets were not then a feature of Southeast Asia's mosques. Multi-roofed structures could not readily accommodate them. Verandas attached to a mosque housed the local solution for alerting the congregation to prayer in the form of a large drum (*bedug*).[17] Cardeel's way out for the five-roofed mosque of Banten was to design a tower for the mosque grounds. This Dutch Muslim perhaps modelled his minaret on the lighthouse from a remembered past in Holland.

16 KITLV 36 A-48, KITLV 27538.

17 The *bedug* became controversial as Indonesians travelled more in Muslim lands of the Middle East and Central Asia. A late nineteenth century fatwa declared it an unacceptable substitute for the muezzin's call to prayer (Kaptein 1997:10).

Through diplomatic exchange the Dutch came into contact with Islamic-inspired art styles such as royal 'golden letters'. These establish the skills of Indigenous illuminators and their participation in a Muslim artistic tradition emanating from Persia and Mughal India. Anonymous Indigenous artists embellished text with borders of flowers and leaves in gold and coloured inks in repeating geometric patterns, and interspersed single flowers within text to indicate new cantos or sections. Elaborate floral frontispieces and colophons begin and end manuscripts.[18]

Little influence of Indonesian art styles or themes is discernible in Dutch paintings from this period, but Asian decorative forms migrated into European-style furniture, such as chairs, armoires, tables and bedsteads. They were made in Batavia workshops, crafted from Coromandel and Indonesian woods, elaborately carved with flowers and foliage by Indian, Sri Lankan and Indigenous woodworkers.[19] Deon Viljoen has traced the journey of one piece, a massive bureau-cabinet with inlays of Asian hardwoods and ivory, he believes was commissioned from Batavia's furniture workshops and Javanese woodcarvers by Hendrik Swellengrebel, governor of the Cape settlement from 1739 to 1751. On repatriation, the cabinet travelled with Swellengrebel from Cape Town to the villa he purchased near Utrecht.[20] In this micro-history we find Javanese decorative forms incorporated into furniture showing the influence of mid-eighteenth century Dutch and German designs. It reminds us that Javanese were involved in the production of other 'European' artworks in Indonesia, for example, mixing paints and stretching canvases in painters' studios.

The Human Likeness: Typologies and Portraits

In the early decades of contact, the Dutch did not travel much inland. Frederik Coyett was the first Dutchman reputed to have seen the Borobudur. That was in 1732. It was the monument's Buddha images that attracted his eye. Borobudur's wall panels, alive with men, women and heavenly beings, framed

18 See examples of royal letters in Gallop (1991, 2011:105–140); *Pameran seabad* 1992: figures 53–55; and examples of decorated manuscripts in Gallop (1995: figures 76–77); and Kumar and McGlynn (1996).

19 See examples in the Rijksmuseum: BK-1994-37, BK-NM-12028-1 and NG-10998. See also De Haan 1922–23:C1-C24.

20 Viljoen 2007:450–459. The inventory of Swellengrebel's possessions shows he also brought to the Netherlands Chinese and Japanese ceramics, lacquered furniture and Indian textiles.

by trees, foliage, birds and monkeys, were possibly not visible to him on account of the dirt, rubble and plants that had accumulated over the centuries. Coyett brought several of the Buddha sculptures back to Batavia and placed them in the garden surrounding the villa he built just before his death in 1736. In time his landed estate passed to Batavia's Chinese community who, also attracted to the Buddha images, transformed Coyett's residence into the Chinese temple and burial ground known as Klenteng Sentiong (now named Vihara Buddhayana).[21]

For much of the seventeenth and early eighteenth centuries, what remains of artists' records, therefore, are harbour city scenes, views of new suburbs outside Batavia's walls, and the country estate of governors-general in Buitenzorg (Bogor). From Johannes Rach (1720-d. Batavia 1783) there are many drawings (ink on paper) of official buildings and residences of the VOC, of churches, tree-lined avenues and public fountains in Batavia, and of the environs of Buitenzorg. Into these settings Rach placed representative groupings: ladies and gentlemen of the VOC elite; Chinese merchants and porters; Indigenous working people in town and fields.[22] Again we see the entwined lives of these recognizable types. J.E. Brandes (1743–1808) has also left many views of Batavia and Batavians, illustrations of Chinese temples, Indonesian troops, Javanese dancers, and a coloured drawing of his own house and extensive grounds.[23]

Into this world of interconnected sea ports the Dutch introduced the portrait. VOC personnel brought likenesses of relatives with them to their postings in Asia; they commissioned portraits of themselves abroad; they had copies made to circulate among family that was scattered across the globe. The VOC commissioned portraits of every governor-general. It became a convention to portray these demi-royals of the Batavian Republic in the ceremonial dress then prevailing in Holland, with rod of office and emblems of the world's first global corporation. The ill-fated governor-general, Adriaan Valckenier (r. 1737–1741), for example, strikes an open pose in happier times, in wig, red velvet jacket and intricately embroidered waistcoat before a drapery pulled aside to allow view of a Dutch ship, the VOC's lifeline.[24] His portraitist, Theodorus

---

21 "Dari Goenoeng Sarie ke Kelenteng Sentiong," *Kompas*, 10 October 2009.

22 See Rach's drawings from 1765–75 in the Rijksmuseum, NG-400-AA, D, I, J, K, L, M, N, O, S, T and Y; and drawings dating from 1770–1772, KITLV 51 C-1-4.

23 198 of Brandes's drawings are examined in De Bruijn and Raben, 2004. The Rijksmuseum holds 600 of his watercolours and drawings. The drawing of his house is catalogued as RM NG-1985-7-2-143.

24 *Portrait of Valckenier*, 1737, by Th. J. Rheen (d. Batavia 1745), RM SK-A-3778, and a copy (by Rheen), RM SK-A-4547.

Justinus Rheen, had begun his artistic career in Amsterdam, but worked in Batavia from 1737.[25] The official portrait of Valckenier's predecessor, Abraham Patras, is also attributed to Rheen.[26]

Few portraits of Dutch women have survived. Of those known, most were of wives and daughters of the VOC's elite. They stand, in the conventions of Dutch portraiture of the age, before balustrades, Grecian urns, columns and draperies. Little anchors them in place. No background scenery, flowers or animals allow viewers to identify a specific Asian locale. Only mountains or ships painted into the background suggest 'the East'. While, at the time, portraits of elite families were commissioned works of known persons, their subjects are now often unknown to us. For example, the child holding a bunch of grapes in a 1663 portrait by Coeman has been identified as Joanna van Riebeeck, younger daughter of the Cape's founder, or alternatively as Johannes van Rees, future brother-in-law to Van Riebeeck.[27] Karel Schoeman attributes another mid-seventeenth century portrait of a young girl to Coeman on the basis of its painterly skill and style, and assumes that Coeman painted her during his layover in Cape Town on the voyage out from Holland in 1664.[28]

The Dutch portrait entered a world with quite different traditions governing the public representation of men and women. In place of the portrait, which explores an individual's character through the face, Indonesian traditions contemporary with the Dutch revealed individuals through family or lineage name and rank titles. Illustrations of heroes and heroines in manuscripts presented them as stylized, *wayang*-like figures.[29] An older, pre-Islamic tradition on Java rendered kings and queens in sculpted conventions of deities.[30] So it was possibly a jarring innovation when portraits of governors-general were presented by VOC officials to sultans and military allies. Did this custom induce a new sense of self? All that can safely be said is that at least one VOC-era sultan

25 Rijksbureau voor Kunsthistorische Documentatie, Entry 66509 (www.rkd.nl).

26 RM SK-A-3777, and copy RM SK-A-4546.

27 J. Taylor (2007:70–76). The portrait in the Rijksmuseum is catalogued RM SK-A-809.

28 Schoeman (2006): back cover. The girl stands brilliant in a red dress before a leafy countryside that is threatened by rain clouds.

29 See, for example, the Princess of Jengala, *Serat Panji Jayakusuma* manuscript, c. 1840, National Library of Indonesia, KBG 139, reproduced in Kumar and McGlynn (1996: x, 182), and *wayang*-like travellers in a two-masted sailing ship (p. 183). Examples from *Serat Damar Wulan* (British Library MSS Jav. 89, early nineteenth century) are in Coster-Wijsman (1953:153–163); Gallop (1991: figure 22); and Gallop (1995: figures 44–46).

30 For example, a late thirteenth century sculpture of Prajnaparamita, Buddhist goddess of wisdom, is thought to represent Ken Dedes, first queen of Singhasari (National Museum of Indonesia).

FIGURE 12.3 *J.J. Coeman,* The Cnoll Family
COLLECTION RIJKSMUSEUM, AMSTERDAM (RM SK-A-4062).

entered into a compact with an unknown Dutch artist to sit for his own portrait. This was Sultan Sayfoeddin of Tidore (r. 1657–1689).[31] Apparently the sultan did not long have the leisure to study his image, for his portrait embarked on its own journeys, eventually coming to Amsterdam's Rijksmuseum by way of Poland and France.

Few named likenesses of Indonesian individuals, as distinct from representative 'types' in paintings and sketches, have survived from the VOC centuries. Léonard Blussé (1997) makes the case that the male servant in Coeman's portrait of the Cnoll family (RM SK-A-4062) is Untung, better known in Indonesian histories as Surapati. Blussé traces his life from Balinese slave, to head of a Balinese militia serving the VOC, to militia commander for Amangkurat II, and finally to ruler of his own realm in East Java. Women servants were favourite subjects for Dutch painters, but their names are unknown. So it is an exception in this genre to be able to identify an Indigenous woman servant in a drawing by Jan

31 Zandvliet (2002): front cover and 121. Some Indonesian dignitaries also sat for their portraits in the later colonial period, even though photographic portraits were then fashionable and cheaper. See portraits of Adipati Mandoera Djajadiningrat, KITLV 5051, and Raden Tumenggung Kartatatanagara, KITLV 37 C-167, both painted in 1846 by Charles William Meredith van de Velde (1818–1898).

Brandes (RM NG-1985-7-2-72). She was his domestic companion and carer for his son. Brandes called her Roosje.

Javanese assistants and apprentices who worked for European artists and draughtsmen became familiar with European art styles. Perhaps some painted parts of large canvases as their counterparts did in Holland. The American medical doctor and naturalist, Thomas Horsfield (1773–1859), who drew Java's antiquities and daily life, trained the Javanese he hired in painting techniques. 'Javanese Procession' by one of his Indigenous staff shows a long line of men and women, drawn in semi-profile, but naturalistically. Some play drums, gongs and a wind instrument; men carry aloft women in ceremonial costume; *payung* bearers follow two men on horseback.[32] From the early nineteenth century some Indian and Chinese artists in Java were painting Indonesia's flora and fauna in the European scientific style that combined precision of observation with beautiful images (Gallop 1995: plates 31–9).

Nature Turned into Landscape

With the expansion of Dutch military and economic control into Java's interior in the nineteenth century, European artists came to know the hinterlands of coastal cities. From 'inner Java' come paintings in which human beings, Dutch and Indonesian, are dwarfed before magnificent mountains, embedded into the natural landscape, rather than imposing themselves on it. Java's mountains fill the canvases Abraham Salm (1801–1876) painted during the 29 years he spent in the Indies. They overwhelm the tiny figures of villagers who celebrate weddings, walk along tracks or sail in small craft.[33] Three Europeans raise their arms to the smoking crater of Gedeh Mountain in a painting by Adrianus Johannes Bik (1790–1872); their Indigenous attendants squat at a distance, apparently indifferent to the natural wonder before them.[34] Jan Simon Gerardus Gramberg (1823–1888) also emphasized the majesty of landscape in which Indonesians pursued their daily lives. In his painting 'Prahu Mountain' (TM 3728-448) a farmer carrying his load trudges through un-noticing; a man in Muslim dress studies his book; Gramberg alone urges us to contemplate the grandeur of nature.[35]

---

32 Gallop (1995): plate 16, early nineteenth-century watercolour.

33 See, for example, *View of Salak Mountain*, KITLV 47 A-72 (painted 1865–1876), *View of Kedong Badak estate and the main road to Bogor*, KITLV 47 D-17 (1872); and *Solo River*, TM 3728–417.

34 *Gedeh Mountain*, c. 1828, TM 3728–528.

35 *Prahu Mountain*, TM 3728–448. See other examples of this genre by F.C. Wilsen (TM 3728–422; KITLV 47 A-78), and by Van de Velde (TM 4108–237).

FIGURE 12.4 *J.S.G. Gramberg,* Prahu Mountain
NATIONAL MUSEUM OF WORLD CULTURES (TM 3728–448).

Even subjects of historical significance could be rendered in soft, romantic style. A painting of the all-Java highway, the Great Post Road, that Governor-General H.W. Daendels (r. 1808–1811) called for and that shifts of Java's corvée labourers laid across mountains, through forests and malarial swamps over 1,000 kilometres at great cost to health and life, exemplifies this style and mentality. The anonymous artist has painted a pleasant scene of smoothly surfaced road winding through stepped rice fields bordered by forest and a long view to distant mountains. Small figures of Indonesian porters and a covered coach are in the foreground (TM 1012–1). Nor did the turmoil, destruction and death toll of the Diponegoro War (1825–1830) in Java intrude into serene scenes of high mountains and luxuriant forests painted by Antoine Auguste Joseph Payen (1792–1853), who first arrived in Java in 1817. In one landscape, painted in 1828, a gentle light bathes the neat house of the assistant-resident of Banyuwangi. His Indigenous servants hold a *payung* above him; messengers who have travelled the Post Road bring him mail (RM SK-A-3452).

This pastoral style continued into the twentieth century, even though Indonesian landscapes had by then been permanently transformed by deforestation, commercial farms, steel bridges, railways, factories, new towns and denser populations. Mountains, rice fields and a huge tree suggest unchanging countryside still overwhelms insignificant humanity in a landscape painted around 1954 by Jan Christiaan Poortenaar (1886–1958). A few Indonesians sit idly in an otherwise empty scene, in which there is no trace of modern Java's motorized bicycles, busy markets and crowds (TM 4754–126). Indonesian artists, such as R. Basuki Abdullah (1915–1993), also painted in this style, called 'Mooi Indië' (beautiful Indies).[36]

Painted History

With paintings of historical events we return to the theme of artists creating images of people and activities they themselves did not witness. Javanese manuscript embellishers had their own traditions of visual history. From the late nineteenth century two history paintings by an unknown Javanese artist show a marriage of Indigenous narrative convention and European elements. Both are watercolours, painted around 1890, of dramatic episodes in the reign of Amangkurat II (r. 1677–1703). The first imagines the execution in 1680 of Prince Trunajaya. At centre the sultan personally stabs the captured rebel with his *kris* 'Noble Blabor' (KITLV 48 M-5). Attendants carry lances and a sword; one holds a *payung* over his sultan. Trunajaya's sisters weep. Dutchmen are incorporated into the composition as witnesses; a Javanese attending them holds aloft a Dutch flag. The second imagines Amangkurat overseeing execution by strangling in 1703 of a faithless wife and her lover (KITLV 48 M-6). Here the sultan is elevated above the scene, seated on a stool; retainers sit cross-legged on the ground. One holds a *payung* over Amangkurat, another two men pay obeisance. Tigers maul the offending wife's women servants who are confined, naked, in a cage.[37] Both historical scenes are set outdoors and framed by trees and birds. While two-dimensional, the figures are drawn with life-like, individual features.

A colonial government building with the royal coat-of-arms, a large Dutch flag and a crowd of uniformed military with lances drawn dominate a history painting by Nicolaas Pieneman (1809–1860). It records Pangeran Diponegoro's enforced departure into exile in Sulawesi (RM SK-A-2238). Diponegoro's capture in 1830 brought five years of warfare to an end. No major obstacle now remained to impede the extension of Dutch power across Java. Willem I of the

36 See for example, *Gedeh Volcano in West Java*, TM 4818–1, painted before 1949.

37 A legend identifies the wife as Raden Ayu Lembah, her lover as Raden Soekra, and names other prominent figures in the painting.

FIGURE 12.5 *Unknown Javanese artist,* Amangkurat oversees execution of wife and lover
UNIVERSITEIT LEIDEN, COLLECTIONS KITLV (48M-6).

Netherlands (r. 1815–40) commissioned Pieneman, who specialized in painting historical subjects and portraits of the famous, to commemorate this turning point in Dutch-Indonesian relations. Pieneman's conception of the traumatic scene was based on a sketch made on the spot by F.V.H.A. de Steurs, aide-de-camp (and son-in-law) to the commanding officer, Baron Hendrik Merkus de Kock. Pieneman may also have consulted a study from 1835 of Diponegoro by the Java-based Bik (TM 1574–32). The defeated prince stands in a circle of light, looking beyond the soldiers across Java's rice fields, past its mountains to the inland sea that will carry him away as directed by De Kock. Pieneman records Diponegoro's self-identification as a Muslim *emir* by the green turban on his head and clothing. Women, in Javanese dress, kneel in despair.

The Javanese artist who excelled in Dutch art techniques, who spent 22 years in Europe, and who was painter by appointment to Willem II (r. 1840–1849), Raden Saleh (c. 1807–1880), also recreated this iconic moment, six years after his own return to Java in 1851.[38] Neither Pieneman nor Saleh acknowledged Diponegoro's illustrious descent with a *payung*. Saleh's rendering of the

38 *The Capture of Prince Diponegoro* by Raden Saleh Syarif Bustaman, 1857, Museum Istana, Jakarta.

scene stages Pieneman's portrayal in reverse. No Dutch flag flies over the crowd. In contrast to the glowing light of Pieneman's painting, Saleh's canvas is dark and gloomy. He depicts Diponegoro's Java by mountains and an arid landscape. The prince's rigid stance is interpreted by Saleh's great-grandnephew, George H. Hundeshagen, as angry and challenging to De Kock.[39] He argues that the painting had a surface meaning apparent to Dutch viewers and a 'clandestine' message of resistance for the Javanese. It is, however, doubtful if many Javanese could have seen Saleh's painting in 1857 (or indeed before 1978) and formed this interpretation as it was sent to the Netherlands upon completion.[40]

In preparation for the painting, Saleh visited sites associated with Diponegoro in Central Java, but in some respects he mirrored Dutch artists of earlier times, for he painted 'Java' from Holland during his long European sojourn that had begun at age fourteen in 1829. Take, for example, his portrait of Governor-General J.C. Baud (r. 1833–1835), painted in Holland in 1835 (RM SK-A-3799). Baud is seated before the drapery and Grecian urn of convention, but against a dawn sky, the Buitenzorg palace and a palm tree, perhaps a scene remembered in Holland from Saleh's youth.[41] Saleh also reflected styles of Dutch contemporaries in painting beautiful Java. In a posthumous portrait of Daendels, also executed in the Netherlands, the governor-general stands against a view of mountain range and forest, before a balustrade. He points to a map, dated 1810, of the projected route of the Post Road. There is a glimpse of the road and tiny figures of workers, but no hint of their wretchedness in its construction (RM SK-A-3790).

Glimpses of Modernity and Emergence of the Individual

Commissioning of official portraits of governors-general outlasted the VOC and continued, through the age of photography, until dissolution of the Netherlands Indies. But photographic images quickly dominate the visual heritage in numbers, range of subjects and geographic scope. Mechanical image-makers imitated painters with views of majestic mountains and beautiful scenery,[42] but they also used the new device to photograph modernity in all its

39 G.H. Hundeshagen, http://www.raden-saleh-id.org/diponegorocapture.html. The article includes a preparatory sketch Raden Saleh made for the painting in 1856.

40 The painting was hung in The Hague palace and later in the Bronbeek Military Veterans' Home. In 1978 the Orange Nassau Foundation presented it to Indonesia.

41 Saleh had studied painting with Payen in Bogor.

42 See, for example, *Salak Mountain viewed from Batutulis*, before 1880, KITLV 26663, and *Confluence of Seumpo and Tjoet Rivers, Lhokseumawe*, 1924, KITLV 18019.

forms. Cameramen followed colonial armies, explorers and entrepreneurs across the archipelago. From the KITLV image archive are photographs of: army troops on the march (5202), individual officers framed in oval portraits (2504), their gravestones in military cemeteries (3415); hillsides stripped of their forest cover (75701); young rubber plantations (75722); railway tracks (27038), road construction (26353), telephone poles (153978), factories (28484), oil rigs (16734), public buildings (34531) and massive banks (35072). Batavia stretches far beyond market and harbour; it encompasses the headquarters of businesses (5224), the building housing the People's Assembly (90168), schools (1400856), parks (100624) and swimming pools (77244).

Dutch, Chinese and Indigenous men were behind the camera. Most famous was the Javanese Kassian Cephas (1845–1912) (Knaap 1999). Cephas's camera recorded Java's antiquities (KITLV 40199), its modern Dutch infrastructure (KITLV 19367), colonial institutions (RM NG-1944-10-13-2), and ceremonies marking the Muslim calendar (KITLV 2351). As official photographer to the sultan, Cephas was admitted into the Yogya *kraton* to record palace dance troupes (KITLV 3905) and make photographic portraits of Hamengkubuwono VII, his queen and crown prince (KITLV 10002, 10003, 10005). These were portraits of people representing office and their office's ideals. For European consumers Cephas photographed young women clad in floral batiks amid abundant nature (KITLV 10727), and sentimental photos of young mothers with babies (KITLV 10728). Cephas also photographed the social life shared by Dutch and Javanese elites, as represented in *Wayang Beber Performance at the House of Wahidin Sudirohusudo, with G.A.J. Hazeu* and *Group Photo on the Back Veranda of Wahidin Sudirohusudo's House, including Hazeu.*[43]

Many photographs in the KITLV archive are of family groups.[44] Alongside the history of building a colony and enmeshing Indonesians into the world economy is the social history of a mixed society that took on local dress forms, spoke Malay, engaged in Indonesian arts and crafts, and incorporated Java's hipped roofs and verandas into their domestic architecture. These family photographs challenge assumptions of segregation in colonial life. They tell us what the demographers and statisticians document, that approximately 80% of the Dutch in the Indies were both Indonesian and Dutch. The immediacy of these photographs causes us to think more deeply about the quality of

43 The catalogue numbers for these two photographs are KITLV 3953 and KITLV 34594. Wahidin (1852–1917), a medical doctor and founder of Budi Utomo, is honoured in Indonesia as a pioneer of nationalism. Hazeu (1870–1929) was a scholar of Indonesian languages, folklore and *wayang*, and advisor for Native and Arab Affairs 1907–1919.

44 See, for example: KITLV 77650, 78505, 86075, 151380 and 503394.

relationships between the people depicted, between them and their environment, and between viewers and viewed. They explain, in the way no written document can, the ambiguities of colonial life, the hostility of nationalists to people who chose the Dutch side of their personal inheritance over (and against) the Indonesian, and the fraught meanings of 'homeland' for Eurasians. Such photographs tell the end to the story that had begun when Indonesians first ferried Dutchmen ashore, a history seventeenth-century Dutch artists had unknowingly recorded in paint.

Well-off Indonesians also took themselves to photographic studios or summoned professional photographers to their residences.[45] They hung framed photographs on their walls for public display and assembled albums that are archives for the emergence of the individual. Photographs allow private study of self, compel acceptance of critical scrutiny by others. Migrated to newspapers, the individual's image is subject to being discarded. Photographs chart the emergence of elite wives and daughters from decorous seclusion on to the public stage and so document the challenge of Dutch female immigrants to existing class, cultural and religious conventions. Photographs commissioned by Indonesian upper classes are a visual, datable history of becoming self-consciously modern, and parallels the new genres of memoir and autobiography that staged and exposed the self to reading publics.

The camera also narrates Indies lives from the perspective of Indigenous partners in colonial rule. In 1924, Raden Toemengoeng Soerjawinoto, *bupati* of Gresik, presented a souvenir album (KITLV 898) to Resident W.P. Hillen and his wife on their departure from Surabaya. Regent Soerjawinoto's album gives us a clear idea of what he anticipated would prompt pleasing memories of Indies service for his Dutch colleague. The very first photograph in the album is of the menu from a dinner, held on 15 July 1924, for the Hillens (94782). It begins with bouillon soup with tomatoes, and promises Veuve Cliquot *demi-sec*. The regent's handwriting labels each photograph's subject and location in Dutch. There are photographs of local scenery, of Soerjawinoto's residence, the Chinese and Arab quarters and cemeteries, plus many of the mosque, grotto and graveyard of Sunan Giri (KITLV 94800–8) and of Njai Ageng Penatij (KITLV 94795), wife of the *patih* of Majapahit who raised the saint. There are photos of Dutch installations, such as water pumping machines, a salt manufacturing site, the Gresik railway station, a village school and a credit bank. In this assemblage we discern the regent's perception of who constituted Surabaya's elite, his pride in Surabaya's importance in the early history of Islam in the region, and his conception of his own role in the modernization processes that Dutch

45 Examples from the KITLV archive are: 4184, 4747, 6354 and 119347.

colonial rule was channelling into Java. There is the claiming of partnership, not subordination.

A private album assembled by an Indigenous family around 1920 in the Moluccas (KITLV 503) conveys the lure of the modern Dutch lifestyle for members of an emerging middle class (H. Schulte Nordholt 2009:105–20). The family appears to be Christian, for there are no visual indications of Islamic culture in the photographs. While photographs in the regent's presentation album show elite life lived in the public gaze, these are mostly photographs of domestic life, taken at home. A family group sits around a table on the veranda (80740). Women wear *kain kebaya*, they hold babies (80733, 80737); a father in singlet and trousers sits in the backyard with children beside a birdcage (80745). Remarkable are views of their dining room with no householder present to distract attention from the table laid with cloth and dinner service, the dresser, cupboard with glass doors and oil lamp (80749). Two unnumbered photographs show a mirror in the hallway and a framed snap of a family group on the wall. In pride of place, photograph number two in the album, is the family's Singer sewing machine operated by a woman seated at a table (80734).

Some photos in the album appear to record service in the lower reaches of colonial administration. It seems the family, in the course of the husband's career, was stationed in an 'Alfoer' village. The album contains photos of curly haired men lined up in a row, naked except for a loincloth (80751), their huts on the edge of a forest (80752, 80753). The album's last photo is of the father in *jas tutup* (collarless, buttoned jacket) and trousers, standing with his wife, dressed in *kain kebaya*, outside their house (80757). Did this family, fully clothed, occupants of the concrete house with glass windows and lace curtains, view Moluccans as 'Others', as uncivilized? Did they see through the eyes of the Dutch? Or did the Dutch perceive such Moluccans through the eyes of other Indonesians? In making them photographable, did the camera impose a way of looking at others? This album, assembled by the family as a record of its own daily life, is suggestive of lifestyle and attitudes of all those Indigenous families whose careers were bound up with the Dutch, but who, not belonging to elite circles, have left no written testimonials of their lives.

Many other photographs in the image archives offer clues to the daily life of thousands of ordinary Indonesians. They show Indigenous men who were lorry drivers (TM 10014093), bicycled through city streets (TM 10014702), were factory workers (TM 60020405) and apprentice draughtsmen (TM 60020323). We see women customers queuing at village credit institutions (TM 10001465), sorting coffee beans (KITLV 26927) and operating sewing machines (KITLV 13110). Alongside these examples of modern lifestyles we see Indies subjects

FIGURE 12.6 *Indonesian family album,* Woman Seated with her Sewing Machine
UNIVERSITEIT LEIDEN, COLLECTIONS KITLV (80734).

(mostly men) becoming Indonesians in photographs that record the development of political life in the late colony.[46]

The earliest moving pictures made in the Indies (1912–1913) also give a visual history of a modernizing colony and new kinds of colonized people. Johann Christian Lamster (1872–1954) was commissioned by Amsterdam's Colonial Institute to film advances achieved by a benign, rational administration in the Indies. Lamster had already spent 15 years in the colony as soldier, civil servant, husband and father. Before his camera women bring their babies to the colonial medical officer (an Indonesian man) for smallpox inoculation, veterinary students practice handling animals, factory workers process sisal into rope, railway employees drive and maintain a steam locomotive. Dutch, Eurasian and Indonesian children spill out of school and climb aboard city trams. Here is Lamster's vision of Holland's colonial subjects: industrious, self-managing, and receptive of the modern.[47]

***

46 See, for example, *Inaugural Meeting of Sarekat Islam in Blitar*, 1914, KITLV 9174; *First Congress of the National Party of Indonesia* (*PNI*) *in Surabaya, 1928*', KITLV 53494; and *Second Congress of the Federation of Political Associations of the Indonesian Nation* (*PPPKI*) *in Surakarta, 1929*, KITLV 53480.

47 Van Dijk et al., (2010). DVD of 15 of Lamster's films.

Visual history complements and enhances the written record. It can challenge received notions by its very focus on the daily round and raise awareness of gaps in knowledge. It may tell us what the written record conceals. As with written records, visual records are accidents of history in what remains and what we select for study. In three centuries of image-making we may note other constraints: the force of artistic conventions, for instance, that indicated European status in Indonesia by urns, draperies, balustrades and heavy winter clothing, or the dictates of early camera technology that controlled stance, place and hour. Here I have drawn attention to the convention of the *payung*. It indicates status in Indonesian hierarchies, but it also shows Dutch people adapting to novel circumstances, how they tried to make themselves understood or asserted rights over Indonesians.

Adopting Braudel's long time span allows us, even from a preliminary survey of images, to chart adaptation and change through mutual interaction in colonial histories. Seascapes give way to urban views and representative 'types'. Landscapes of sublime nature and insignificant humanity are, seen through the camera's lens, denuded hills and agricultural estates where men and women toil. Portraits of elites give way to photos of commoners. Photographs of officials tell us of the alliance between Dutch and Indonesian elites that made colonial rule possible. Photographs of families and individuals signal a modern consciousness of self.

Reliability of the visual record is a constant concern for historical study. We have noted paintings of places and people imagined by artists who themselves saw neither the sites nor individuals that filled their canvases, but relied on the information of others who had. And we have noted examples of artists who did work in the Indies but chose, in painting nature, not to see the transformations before them that modern technology and increasing population were bringing to Java's countryside. The camera allowed a wider range of subjects and a greater spread of social classes, situations and moods to be recorded. Through its unique emphasis on the human face, the camera hints of relationships and complications behind the staged pieces, the set expressions or pretence of unrehearsed moment.

Through seven decades of mechanically produced images we discern the desire of camera owners to record modern life. Thus we learn that Indonesian Everyman and Everywoman embraced new technology. They became drivers of lorries, operators of factory machinery. They did not shun the new. We see adopting and adapting in Lamster's film clip of the Indonesian paramedic vaccinating village babies. An intrusive medical procedure is turned into public ceremony. Film director, vaccinator and village authorities colluded in a 'performance'. But we also learn that 'housewives', in submitting their babies to this new medical practice, were also participants and partners in colonial modernity.

We see altered perspectives most strikingly in the newly acquired habit of Indonesians viewing themselves. The camera has never left Indonesian hands; it is an indelible part of contemporary Indonesian culture. Alongside letters and autobiography, photographs document the emergence of the named individual in Indonesian life. This novel way of understanding self and society began when Sultan Sayfoeddin and Roosje studied their portraits.

Artists and art forms cross cultural borders. We find a common delight in lovely things, a shared impulse for learning, and a medium revelatory of the hidden, the voiceless and the byways of history.

List of Writings of Peter Boomgaard

Forthcoming

Books

Histories of the Forests of Java, 1500–1950 [working title].

Edited Volumes

With Raquel Reyes, *The Senses in Southeast Asian Pasts: Forms of Perception across two Millennia*. London/New York: Routledge.

Articles

With Pieter M. Kroonenberg, "Rice, Sugar, and Livestock in Java, 1820–1940: Geertz's Agricultural Involution 50 years on." In *Rice: Global Networks and New Histories*, edited by Francesca Bray et al. Cambridge: Cambridge University Press.

"Environmental Histories of Southeast Asia." In *Handbook of the Environment in Southeast Asia*, edited by Philip Hirsch. London/New York: Routledge.

"Smells and Sounds of the City: Batavia, Java, in the 17th and 18th Centuries." In *The Senses in Southeast Asian Pasts: Forms of Perception across two Millennia*, edited by Raquel Reyes and Peter Boomgaard. London/New York: Routledge.

Published

Books

Southeast Asia: An Environmental History. Santa Barbara: ABC-CLIO, 2007, 377 pp.

Anak Jajahan Belanda: Sejarah Sosial dan Ekonomi Jawa 1795–1880. Jakarta: KITLV Jakarta/Djambatan, 2004, 413 pp. [Translation of *Children of the Colonial State*, Amsterdam: VU University Press.]

Frontiers of Fear: Tigers and People in the Malay World, 1600–1950. New Haven/London: Yale University Press, 2001, 306 pp.

With Janneke van Dijk. *Het Indië Boek*. Zwolle: Waanders, 2001, 512 pp.

With the assistance of R. de Bakker. *Forests and Forestry 1823–1941*. Amsterdam: KIT, 1996, 184 pp. [Changing Economy in Indonesia 16.]

With A.J. Gooszen. *Population Trends 1795–1942*. Amsterdam: KIT, 1991, 256 pp. [Changing Economy in Indonesia 11.]

With J.L. van Zanden. *Food Crops and Arable Lands, Java 1815–1942*. Amsterdam: KIT, 1990, 144 pp. [Changing Economy in Indonesia 10.]

 | DOI 10.1163/9789004288058_014

Between Sovereign Domain and Servile Tenure: The Development of Rights to Land in Java, 1780–1870. Amsterdam: Free University Press, 1989, 55 pp. [Comparative Asian Studies 4.]

Children of the Colonial State: Population Growth and Economic Development in Java: 1795–1880. Amsterdam: VU University Press, 1989, 247 pp. [CASA Monographs 1.]

Edited Volumes

Empire and Science in the Making – Dutch Colonial Scholarship in Comparative Global Perspective, 1760–1830. New York/Houndmills, Basingstoke: Palgrave Macmillan, 2013, 320 pp.

Special Issue: "Long-Term Changes in Land-Tenure Arrangements in Pre-Modern and Early-Modern Southeast Asia." *Journal of the Economic and Social History of the Orient* 54 (4), 2011, pp. 447–519.

With Marjolein 't Hart. *Globalization, Environmental Change and Social History*. Cambridge: Cambridge University Press, 2010, 233 pp. [Also published as a Special Issue of *International Review of Social History* (volume 55, issue S18, 2010).]

With David Henley. *Credit and Debt in Southeast Asia, 860–1930: From Peonage to Pawnshop, from Kongsi to Cooperative*. Singapore: ISEAS, 2009, 195 pp.

With Dick Kooiman and Henk Schulte Nordholt. *Linking Destinies: Trade, Towns and Kin in Asian History*. Leiden: KITLV, 2008, 277 pp.

A World of Water: Rain, Rivers and Seas in Southeast Asian Histories. Leiden: KITLV Press, 2007, 368 pp.

With Greg Bankoff. *A History of Natural Resources in Asia: The Wealth of Nature*. New York/Houndmills, Basingstoke: Palgrave Macmillan, 2007, 288 pp.

With David Henley and Manon Osseweijer. *Muddied Waters: Historical and Contemporary Perspectives on Management of Forests and Fisheries in Island Southeast Asia*. Leiden: KITLV Press, 2005, 418 pp.

With David Henley. *Smallholders and Stockbreeders: Histories of Foodcrop and Livestock Farming in Southeast Asia*. Leiden: KITLV Press, 2004, 344 pp.

With Ian Brown. *Weathering the Storm: The Economies of Southeast Asia in the 1930s Depression*. Singapore: ISEAS, Leiden: KITLV Press, 2000, 332 pp.

With Freek Colombijn and David Henley. *Paper Landscapes: Explorations in the Environmental History of Indonesia*. Leiden: KITLV Press, 1997, 424 pp.

With Harry A. Poeze and Gerard Termorshuizen. *God in Indië: Bekeringsverhalen uit de Negentiende Eeuw*. Leiden: KITLV Uitgeverij, 1997, 156 pp.

With Rosalia Sciortino and Inez Smyth. *Health Care in Java: Past and Present*. Leiden: KITLV Press, 1996, 197 pp. [KITLV Proceedings 3.]

With Harry Poeze and Gerard Termorshuizen. *Aangeraakt door Insulinde*. Leiden: KITLV Uitgeverij, 1992, 143 pp.

With Paul Alexander and Ben White. *In the Shadow of Agriculture: Non-farm Activities in the Javanese Economy, Past and Present*. Amsterdam: KIT, 1991, 154 pp.

The Colonial Past: Dutch Sources on Indonesian History. Amsterdam: KIT, 1991, 63 pp. [Bulletin 323, Royal Tropical Institute.]

With L. Noordegraaf, H. De Vries, and W.M. Zappey. *Exercities in ons Verleden: Twaalf Opstellen over de Economische en Sociale Geschiedenis van Nederland en Koloniën, 1800–1950: Aangeboden aan Prof. W.J. Wieringa*. Assen: Van Gorcum, 1981, 254 pp.

Articles in Refereed Journals

"Long-Term Changes in Land-Tenure Arrangements in Pre-Modern and Early-Modern Southeast Asia: An Introduction." *Journal of the Economic and Social History of the Orient* 54 (4) (2011): 447–454.

"Land Rights and the Environment in the Indonesian Archipelago, 800–1950." *Journal of Economic and Social History of the Orient* 54 (4) (2011): 478–496.

With Marjolein 't Hart. Special Issue 18: "Globalization, Environmental Change and Social History: An Introduction." *International Review of Social History* 55 (2010): 1–26. [Edited by Peter Boomgaard and Marjolein 't Hart.]

"Droefenis en Duurzaamheid: Beheer en Exploitatie van de Bossen op Java onder Daendels (1808–1810)." *Jaarboek voor Ecologische Geschiedenis 2009* (2010): 59–84.

"Labour, Land, and Capital Markets in Early Modern Southeast Asia from the Fifteenth to the Nineteenth Century." *Continuity and Change* 24 (1) (2009): 55–78.

"Technologies of a Trading Empire: Dutch Introduction of Water- and Windmills in Early-modern Asia, 1650s–1800." *History and Technology* 24 (1) (2008): 41–59.

"Syphilis, Gonorrhoea, Leprosy and Yaws in the Indonesian Archipelago, 1500–1950." *Manusya, Journal of Humanities* 14 (2007): 20–41.

"Zeven Eeuwen Confrontaties tussen Mens en Tijger." *Leidschrift* 21 (1) (2006): 73–88.

"Tijgerstekerijen en Tijger-Buffelgevechten op Java, 1620–1906." *Indische Letteren* 21 (1) (2006): 45–56.

"The Making and Unmaking of Tropical Science: Dutch Research on Indonesia, 1600–2000." *Bijdragen tot de Taal-, Land- en Volkenkunde* 162 (2/3) (2006): 191–217.

"Bridewealth and Birth Control: Low Fertility in the Indonesian Archipelago, 1500–1900." *Population and Development Review* 29 (2) (2003): 197–214.

"In the Shadow of Rice: Roots and Tubers in Indonesian History, 1500–1950." *Agricultural History* 77 (4) (2003): 582–610.

"Human Capital, Slavery and Low Rates of Economic and Population Growth in Indonesia, 1600–1910." *Slavery & Abolition: A Journal of Slave and Post-Slave Studies* 24 (2) (2003): 83–96.

"Smallpox, Vaccination, and the 'Pax Neerlandica': Indonesia, 1550–1930." *Bijdragen tot de Taal-, Land- en Volkenkunde* 159 (4) (2003): 590–617.

"From Subsistence Crises to Business Cycle Depressions, Indonesia 1800–1940." *Itinerario* 26 (3–4) (2002): 35–49.

"Oriental Nature, its Friends and its Enemies: Conservation of Nature in Late-Colonial Indonesia, 1889–1949." *Environment and History* 5 (3) (1999): 257–292.

"Het Javaanse Boerenbedrijf, 1900–1940." *NEHA-Jaarboek voor Economische, Bedrijfs- en Techniekgeschiedenis* 62 (1999): 173–185.

"Environmental Impact of the European Presence in Southeast Asia, 17th-19th centuries." *Illes i Imperis* (1) (1998): 21–35.

"The Miracle of the Second Flowering: Recent Developments in Southeast Asian Studies in Europe." *Southeast Asian Studies* 35 (1) (1997): 164–169.

"Ten geleide." *Spiegel Historiael* 32 (10/11) (1997): 414–417. [Special issue: "People and Nature in Indonesia: Environmental History 1500-present."]

"Gevolgen van de Introductie van Nieuwe Landbouwgewassen (1600–1900)." *Spiegel Historiael* 32 (10/11) (1997): 418–423. [Special issue: "People and Nature in Indonesia: Environmental History 1500-present."]

"Death to the Tiger! The Development of Tiger and Leopard Rituals in Java, 1605–1906." *South East Asia Research* 2 (2) (1994): 141–175.

"Historical Studies in 150 volumes of Bijdragen." *Bijdragen tot de Taal-, Land- en Volkenkunde* 150 (4) (1994): 685–702.

"The Development of Colonial Health Care in Java: An Exploratory Introduction." *Bijdragen tot de Taal-, Land- en Volkenkunde* 149 (1) (1993): 77–93.

"Forest Management and Exploitation in Colonial Java, 1677–1897." *Forest and Conservation History* 36 (1) (1992): 4–14.

"The Tropical Rain Forests of Suriname: Exploitation and Management 1600–1975." *New West Indian Guide/Nieuwe West-Indische Gids* 66 (3–4) (1992): 207–235.

"The Javanese Village as a Cheshire Cat: The Java Debate against a European and Latin American Background." *The Journal of Peasant Studies* 18 (2) (1991): 288–304.

"Why Work for Wages? Free Labour in Java, 1600–1900." *Economic and Social History in the Netherlands* 2 (1990): 37–56.

"'Rough Music' outside Europe: the case of Nineteenth-century Java." *Economic and Social History in the Netherlands* 1 (1989): 141–148.

With Gert J. Oostindie. "Changing Sugar Technology and the Labour Nexus: The Caribbean, 1750–1900." *Nieuwe West-Indische Gids/New West Indian Guide* 63 (1/2) (1989): 5–15.

"De Welvaartspolitiek in Nederlands-Indië (1900–1942)." *Spiegel Historiael* 22 (9) (1987): 382–387.

"Multiplying Masses: Nineteenth Century Population Growth in India and Indonesia." *Itinerario* 11 (1) (1987): 135–148.

"Buitenzorg in 1805: The Role of Money and Credit in a Colonial Frontier Society." *Modern Asian Studies* 20 (1) (1986): 33–58.

With W.L. Korthals Altes. "Indonesië's Verleden in Cijfers: Het Project 'Changing Economy in Indonesia'" *Orion* 2 (5) (1986): 35–37.

"The Welfare Services in Indonesia, 1900–1942." *Itinerario* 10 (1) (1986): 57–81.

"Arbeidersgeschiedenis in Nederland: Een Proeve tot een Sociale Psychologie." *Tijdschrift voor Sociale Geschiedenis* 10 (4) (1984): 368–379. [This article was awarded a prize in 1985 by the Sociaal Wetenschappelijke Raad of the Koninklijke Nederlandse Akademie van Wetenschappen.]

"Population Growth and Welfare in Java (1800–1942)." *Kabar Seberang* 13/14 (1984): 82–102.

"Surinam Plantations in Dutch Archives." *Itinerario* 6 (1) (1982): 121–126.

"Female Labour and Population Growth in Nineteenth Century Java." *Review of Indonesian and Malayan Affairs* 15 (2) (1981): 1–34.

"Bevolkingsgroei en Welvaart op Java, 1800–1942." *Intermediair* 15 (32) (1979): 25–35.

"Revolutie als Theoretische Categorie." *Lampas* 11 (2) (1978): 70–95.

"De Havenstaking van 1946 in Amsterdam en Rotterdam: Confrontatie met Conflict en Stakingstheorieën." *Economisch- en Sociaal-Historisch Jaarboek* 40 (1977): 242–312.

"¿Una revolución inevitable? México 1876–1911." *Colmena Universitaria* 3 (1975): 12–38. [Univ. de Guanajuato, Mexico.]

Articles in Edited Volumes

"The Demographic History of Southeast Asia in the Twentieth Century." In *Histories of Health in Southeast Asia: Perspectives on the Long Twentieth Century*, edited by Tim Harper and Sunil S. Amrith, 87–98. Bloomington: Indiana University Press, 2014.

"From the Mundane to the Sublime: Science, Empire, and the Enlightenment (1760s–1820s)." In *Empire and Science in the Making – Dutch Colonial Scholarship in Comparative Global Perspective, 1760–1830*, edited by Peter Boomgaard, 1–37. New York/Houndmills, Basingstoke: Palgrave Macmillan, 2013.

"For the Common Good: Dutch Institutions and Western Scholarship on Indonesia around 1800." In *Empire and Science in the Making – Dutch Colonial Scholarship in Comparative Global Perspective, 1760–1830*, edited by Peter Boomgaard, 135–164. New York/Houndmills, Basingstoke: Palgrave Macmillan, 2013.

"Population Growth and Environmental Change: A Two-Track Model." In *The Routledge Handbook of Southeast Asian History*, edited by Norman Owen, 133–143. London/New York: Routledge, 2013.

"Southeast Asia in Global Environmental History." In *A Companion to Global Environmental History*, edited by J.R. McNeill and Erin Stewart Mauldin, 81–95. Malden: Wiley-Blackwell, 2012.

"Male-Male Sex, Bestiality and Incest in the Early-Modern Indonesian Archipelago: Perceptions and Penalties." In *Sexual Diversity in Asia, c. 600–1950*, edited by Raquel A.G. Reyes and William G. Clarence-Smith, 141–160. London/New York: Routledge, 2012.

"Shame, Savings, and Colonial Expansion: Patterns of Migration and Family Structure in Sumatra and Java, 1800–1950." In *Southeast Asian Historiography: Unravelling the Myths: Essays in Honour of Barend Jan Terwiel*, edited by Volker Grabowsky, 236–247. Bangkok: River Books, 2011.

"Tigers and People in the Malay World: Four Centuries of Confrontation and Co-existence in Comparative Asian Perspective." In *Tigers of the World: The Science, Politics and Conservation of Panthera tigris*, edited by Ronald Tilson and Philip J. Nyhus, 349–356. London: Academic Press, 2010.

"'Following the Debt': Credit and Debt in Southeast Asian Legal Theory and Practice, 1400–1800." In *Credit and Debt in Southeast Asia, 860–1930: From Peonage to Pawnshop, from Kongsi to Cooperative*, edited by David Henley and Peter Boomgaard, 61–79. Singapore: ISEAS, 2009.

"Writing Medical Histories of Southeast Asia: Comparative Approaches to Disease, Health and Healing." In *Proceedings First International Conference on the History of Medicine in the Philippines, January 9 & 10, 2007, Manila, Philippines*, edited by Angel Aparicio, 1–16. Manila: Miguel de Benavides Library, University of Santo Tomas, 2008.

With Henk Schulte Nordholt. "Connecting People, Places, and Commodities." In *Linking Destinies: Trade, Towns and Kin in Asian History*, edited by Peter Boomgaard, Dick Kooiman and Henk Schulte Nordholt, 1–12. Leiden: KITLV, 2008.

"Early Globalization: Cowries as Currency, 600 BCE-1900." In *Linking Destinies: Trade, Towns and Kin in Asian History*, edited by Peter Boomgaard, Dick Kooiman and Henk Schulte Nordholt, 13–28. Leiden: KITLV, 2008.

"Horse Breeding, Long-Distance Horse Trading and Royal Courts in Indonesian History, 1500–1900." In *Breeds of Empire: The "Invention" of the Horse in Southeast Asia and Southern Africa, 1500–1950*, edited by Greg Bankoff and Sandra Swart, 33–49. Copenhagen: NIAS, 2007.

"In a State of Flux: Water as a Deadly and a Life-Giving Force in Southeast Asia." In *A World of Water: Rain, Rivers and Seas in Southeast Asian Histories*, edited by Peter Boomgaard, 1–23. Leiden: KITLV Press, 2007.

With Greg Bankoff. "Introduction: Natural Resources and the Shape of Asian History, 1500–2000." In *A History of Natural Resources in Asia: The Wealth of Nature*, edited by Greg Bankoff and Peter Boomgaard, 1–17. New York/Houndmills, Basingstoke: Palgrave Macmillan, 2007.

"From Riches to Rags? Rice Production and Trade in Asia, Particularly Indonesia, 1500–1950." In *A History of Natural Resources in Asia: The Wealth of Nature*, edited by Greg

Bankoff and Peter Boomgaard, 185–203. New York/Houndmills, Basingstoke: Palgrave Macmillan, 2007.

"Cockfights and Quailfights in Indonesia, 800–1950: Male versus Female Aggression." In *Les Messagers divins: Aspects esthétiques et symboliques des Oiseaux en Asia du Sud-Est/Divine Messengers; Bird Symbolism and Aesthetics in Southeast Asia*, edited by Pierre Le Roux and Bernard Sellato, 119–160. [n.p.]: Éditions Connaissances et Savoir/SevenOrients/IRASEC, 2006.

"Products and People of the Sea in and around the Indonesian Archipelago, 900–1900." In *Muddied Waters: Historical and Contemporary Perspectives on Management of Forests and Fisheries in Island Southeast Asia*, edited by Peter Boomgaard, David Henley, and Manon Osseweijer, 97–120. Leiden: KITLV Press, 2005.

"The Long Goodbye? Trends in Forest Exploitation in the Indonesian Archipelago, 1600–2000." In: *Muddied waters: Historical and Contemporary Perspectives on Management of Forests and Fisheries in Island Southeast Asia*, edited by Peter Boomgaard, David Henley, and Manon Osseweijer, 211–234. Leiden: KITLV Press, 2005.

"'Primitive' Tiger Hunters in Indonesia and Malaysia, 1800–1950." In *Wildlife in Asia: Cultural Perspectives*, edited by John Knight, 185–206. London/New York: RoutledgeCurzon, 2004.

"Steden met een Verleden." In *Andere Verhalen over Azië en Onderzoek: Aangeboden aan Ot van den Muijzenberg bij zijn Afscheid als Hoogleraar Sociologie en Moderne Geschiedenis van Niet-westerse Samenlevingen, in het Bijzonder van Zuid- en Zuidoost-Azië, aan de Universiteit van Amsterdam*, edited by Rosanne Rutten and Loes Schenk-Sandbergen, 114–118. Amsterdam: Het Spinhuis, 2004.

"Demographic Transition in Southeast Asia." In *Southeast Asia: A Historical Encyclopedia, from Angkor Wat to East Timor*, edited by Ooi Keat Gin, 414–418. Vol. 1. Santa Barbara: ABC-CLIO, 2004.

With David Henley. "Agriculture and Livestock Histories of Southeast Asia: Approaches and Perspectives." In *Smallholders and Stockbreeders: Histories of Foodcrop and Livestock Farming in Southeast Asia*, edited by Peter Boomgaard and David Henley, 1–16. Leiden: KITLV Press, 2004.

"Horses, Horse-Trading, and Royal Courts in Indonesian History, 1500–1900." In *Smallholders and Stockbreeders: Histories of Foodcrop and Livestock Farming in Southeast Asia*, edited by Peter Boomgaard and David Henley, 211–232. Leiden: KITLV Press, 2004.

"The Age of the Buffalo and the Dawn of the Cattle Era in Indonesia, 1500–1850." In *Smallholders and Stockbreeders: Histories of Foodcrop and Livestock Farming in Southeast Asia*, edited by Peter Boomgaard and David Henley, 257–282. Leiden: KITLV Press, 2004.

"Human Capital, Slavery and Low Rates of Economic and Population Growth in Indonesia, 1600–1910." In *The Structure of Slavery in Indian Ocean Africa and Asia*, edited by Gwyn Campbell, 83–96. London/Portland: Frank Cass, 2004.

"The High Sanctuary: Local Perceptions of Mountains in Indonesia, 1750–2000." In *Framing Indonesian Realities: Essays in Symbolic Anthropology in Honour of Reimar Schefold*, edited by Peter Nas, Gerard Persoon and Rivke Jaffe, 295–314. Leiden: KITLV Press, 2003.

With Ben White and Milan Titus. "The Experience of Crisis in Indonesia: Comparative, local and historical dimensions." In *Indonesia in Search of Transition*, edited by Henk Schulte Nordholt and Irwan Abdullah, 149–176. Yogyakarta: Pustaka Pelajar, 2002.

"Jagung dan Tembakau di Dataran Tinggi Indonesia, 1600–1940." In *Proses Transformasi Daerah Pedalaman di Indonesia*, edited by Tania Murray Li, 75–124. Jakarta: Obor, 2002.

"Crisis Mortality in Seventeenth-Century Indonesia." In *Asian Population History*, edited by Ts'ui-jung Liu et al., 191–220. Oxford: Oxford University Press, 2001.

"A Bird's-Eye View of Economic and Social Development in the District of Comal, 1750–1940." In *Beneath the Smoke of the Sugar-Mill: Javanese Coastal Communities during the Twentieth Century*, edited by Hiroyoshi Kano, Frans Hüsken and Djoko Suryo, 9–37. Yogyakarta: Akatiga & Gadjah Mada University Press, 2001.

"Changing Economic Policy." In *South East Asia: Colonial History. Volume III: High Imperialism (1890s–1930s)*, edited by Paul H. Kratoska, 77–94. London/New York: Routledge, 2001.

"Peper, Goud en Menseneterij." In *Oosterse omzwervingen: Klassieke teksten over Indonesië uit Oost en West*, edited by Harry A. Poeze, 187–190. Leiden: KITLV Uitgeverij, 2000.

With Ian Brown. "The Economies of Southeast Asia in the 1930s Depression: An introduction." In *Weathering the Storm: The Economies of Southeast Asia in the 1930s Depression*, edited by Peter Boomgaard and Ian Brown, 1–19. Singapore/Leiden: ISEAS/KITLV Press, 2000.

"Surviving the Slump: Development in Real income During the Depression of the 1930s in Indonesia, particularly Java." In *Weathering the Storm: The Economies of Southeast Asia in the 1930s Depression*, edited by Peter Boomgaard and Ian Brown, 23–52. Singapore/Leiden: ISEAS/KITLV Press, 2000.

"Kekayaan-kekayaan Haram: Perkembangan Ekonomi dan Perubahan Sikap terhadap Uang dan Kekayaan Seperti Tercermin dalam Kepercayaan Jawa Populer." In *Sejarah Ekonomi Modern Indonesia: Bergabai Tantangan Baru*, edited by J. Thomas Lindblad, 281–307. Jakarta: LP3ES, 2000.

"Maize and Tobacco in Upland Indonesia, 1600–1940." In *Transforming the Indonesian Uplands: Marginality, Power and Production*, edited by Tania Murray Li, 45–78. Amsterdam: Harwood Academic Publishers, 1999.

"Het Regenwoud en de Ontwikkeling van de Landbouw." In *Geschiedenis van Indonesië: Land, Volk en Cultuur. Deel 2, Vroeg-moderne Geschiedenis*, edited by A. Reid, 166–167. Abcoude: Uniepers, 1998.

"Voedsel, Feest en Hongersnood." In *Geschiedenis van Indonesië: Land, Volk en Cultuur. Deel 2, Vroeg-moderne Geschiedenis*, edited by A. Reid, 168–169. Abcoude: Uniepers, 1998.

"Nederlands Bestuur en de Bevolkingsexplosie." In *Geschiedenis van Indonesië: Land, Volk en Cultuur. Deel 2, Vroeg-moderne Geschiedenis*, edited by A. Reid, 274–275. Abcoude: Uniepers, 1998.

"The VOC Trade in Forest Products in the 17th Century." In *Nature and the Orient: The Environmental History of South and Southeast Asia*, edited by R.H. Grove, V. Damodaran and S. Sangwan, 375–395. Oxford University Press, 1998.

"Peningkatan yang Sia-sia? Dampak Langkah-langkah Kesejahteraan pada masa Kolonial lanjut di Indonesia." In *Pembangunan dan Kesejahteraan Sosial: Indonesia di bawah Orde Baru*, edited by F. Hüsken, M. Rutten and J.-P. Dirkse, 291–298. Jakarta: Gramedia Widiasarana Indonesia (Grasindo): Perwakilan Koninklijk Instituut voor Taal-, Land- en Volkenkunde, 1997.

"Forest Management and Exploitation in Colonial Java, 1677–1897." In *Agriculture, Resource Exploitation, and Environmental Change*, edited by H. Wheatley, 291–311. Aldershot/Brookfield: Variorum, 1997. [An Expanding World; The European Impact on World History 1450–1800, Vol. 17.]

"Introducing Environmental Histories of Indonesia." In *Paper Landscapes: Explorations in the Environmental History of Indonesia*, edited by P. Boomgaard, F. Colombijn and D. Henley, 1–26. Leiden: KITLV Press, 1997.

"Hunting and Trapping in the Indonesian Archipelago, 1500–1950." In *Paper Landscapes: Explorations in the Environmental History of Indonesia*, edited by P. Boomgaard, F. Colombijn and D. Henley, 185–213. Leiden: KITLV Press, 1997.

"Selayang Pandang Perkembangan Ekonomi dan Sosial Daerah Comal Periode 1750–1940." In *Di bawah Asap Pabrik Gula: Masyarakat Desa di Pesisir Jawa Sepanjang Abad ke-20*, edited by H. Kano, F. Hüsken and Djoko Surjo, 11–40. Yogyakarta: Gadjah Mada University Press, 1996.

"Dutch Medicine in Asia, 1600–1900." In *Warm Climates and Western Medicine: The Emergence of Tropical Medicine, 1500–1900*, edited by D. Arnold, 42–64. Amsterdam/Atlanta: Rodopi, 1996.

"Geld, Krediet, Rente en Europeanen in Zuid- en Zuidoost-Azië in de Zeventiende Eeuw." In *Kapitaal, Ondernemerschap en Beleid: Studies over Economie en Politiek in Nederland, Europa en Azië van 1500 tot Heden. Afscheidsbundel voor Prof. dr. P.W. Klein*, edited by C.A. Davids, W. Fritschy and L.A. van der Valk, 483–510. Amsterdam: NEHA, 1996.

"Forests and the Advance of Agriculture." In *Indonesian Heritage: Early Modern History*, edited by A. Reid, 14–15. Singapore: Archipelago Press, 1996.

"Food, Feasting and Famines." In *Indonesian Heritage: Early Modern History*, edited by A. Reid, 15–16. Singapore: Archipelago Press, 1996.

"Dutch Rule and the Population Explosion." In *Indonesian Heritage: Early Modern History*, edited by A. Reid, 122–123. Singapore: Archipelago Press, 1996.

"Sacred Trees and Haunted Forests in Indonesia Particularly Java, Nineteenth and Twentieth Centuries." In *Asian Perceptions of Nature: A Critical Approach*, edited by O. Bruun and A. Kalland, 47–62. Richmond, Surrey: Curzon Press, 1995.

"Colonial Forest Policy in Java in Transition, 1865–1916." In *The Late Colonial State in Indonesia: Political and Economic Foundations of the Netherlands Indies 1880–1942*, edited by R. Cribb, 117–137. Leiden: KITLV Press, 1994.

"De Bibliotheek, de Indonesië-studies en de Overheid." In *Arends Repertorium. De Centrale Bibliotheek van 1971 tot 1993*, 43–45. Amsterdam: KIT Press, 1993. [Bulletin 327, Royal Tropical Institute.]

"Economic Growth in Indonesia, 500–1990." In *Explaining Economic Growth*, edited by A. Szirmai, B. van Ark and D. Pilat, 195–216. Amsterdam: Elsevier Science Publishers, 1993.

"Upliftment Down the Drain?: Effects of Welfare Measures in Late Colonial Indonesia." In *Development and Social Welfare: Indonesia's Experiences under the New Order*, edited by J.-P. Dirkse, F. Hüsken and M. Rutten, 247–253. Leiden: KITLV Press, 1993.

"Illicit Riches: Economic Development and Changing Attitudes towards Money and Wealth as reflected in Javanese Popular Belief." In *New Challenges in the Modern Economic History of Indonesia*, edited by J.Th. Lindblad, 197–215. Leiden: Programme of Indonesian Studies, 1993.

"Protection de la Nature en Indonésie pendant la Fin de la Période Coloniale (1889–1949)." In *Revue Française d'Histoire d'Outre-Mer* 80 (299) (1993): 307–344. [Also in J. Pouchepadass (ed.), *Colonisation et Environnement*, 307–344. Paris: Société Française d'Histoire d'Outre-Mer, 1993.]

"De eerste Antropoloog." In *Aangeraakt door Insulinde*, edited by P. Boomgaard, H. Poeze en G. Termorshuizen, 9–13. Leiden: KITLV Uitgeverij, 1992.

"Exploitation and Management of the Surinam Forests 1600–1975." In *Changing Tropical Forests. Historical Perspectives on Today's Challenges in Central & South America*, edited by H.K. Steen and R.P. Tucker, 252–264. [n.p.]: Forest History Society, 1992. [Proceedings of a conference sponsored by the Forest History Society and IUFRO Forest History Group.]

"Sacred Trees and Haunted Forests – Indonesia, particularly Java, 19th and 20th Centuries." In *Asian Perceptions of Nature. Papers presented at a Workshop, NIAS, Copenhagen, Denmark, October 1991*, edited by O. Bruun and A. Kalland, 39–53. Copenhagen: NIAS, 1992. [Nordic Proceedings in Asian Studies, No. 3.]

"Changing Economy in Indonesia." In *The Colonial Past: Dutch Sources on Indonesian History*, edited by P. Boomgaard, 54–61. Amsterdam: KIT, 1991. [Bulletin 323, Royal Tropical Institute.]

With Susan Legêne. "Introduction." In *The Colonial Past: Dutch Sources on Indonesian History*, edited by P. Boomgaard, 7–11. Amsterdam: KIT, 1991. [Bulletin 323, Royal Tropical Institute.]

"The Non-agricultural Side of an Agricultural Economy, Java, 1500–1900." In *In the Shadow of Agriculture: Non-farm Activities in the Javanese Economy, Past and Present*, edited by P. Alexander, P. Boomgaard and B. White, 14–40. Amsterdam: KIT, 1991.

"Changing Measures and Measuring Changes: Regional Agricultural Growth in Java, 1815–1875." In *Indonesian Economic History in the Dutch Colonial Era*, edited by A. Booth, W.J. O'Malley and A. Weidemann, 111–135. New Haven: Yale University Press, 1990. [Monograph series 35, Yale University Southeast Asia Studies.]

"Klimaat, Krokodillen en Kapitalisme: Economische en Sociale Geschiedenis van Zuidoost-Azië 1500 tot Heden." In *Onderzoek in Zuidoost-Azië: Agenda's voor de jaren Negentig*, edited by H. Schulte Nordholt, 51–62. Leiden: Vakgroep Talen en Culturen van Zuidoost-Azië en Oceanië, RUL, 1990. [Semaian 3.]

"The Javanese Rice Economy 800–1800." In *Economic and Demographic Development in Rice Producing Societies: Some Aspects of East Asian Economic History (1500–1900)*, edited by A. Hayami and Y. Tsubouchi, 142–161. Leuven: Leuven University Press, 1990. [Studies in Social and Economic History 6.]

"The Javanese Rice Economy 800–1800." In *Economic and Demographic Development in Rice Producing Societies: Some aspects of East Asian Economic History (1500–1900)*, edited by A. Hayami and Y. Tsubouchi, 317–344. Tokyo: [Keio University] 1989.

"Pokken en Vaccinatie op Java, 1780–1860: Medische Gegevens als Bron voor de Demografische Geschiedenis." In *Nederlandse Geneeskunde in de Indische Archipel 1861–1942*, edited by A.M. Luyendijk-Elshout et al., 115–127. Amsterdam/Atlanta: Rodopi, 1989.

"Smallpox and Vaccination on Java, 1780–1860: Medical Data as a Source for Demographic History." In *Dutch Medicine in the Malay Archipelago, 1816–1942*, edited by A.M. Luyendijk-Elshout et al., 119–131. Amsterdam/Atlanta: Rodopi, 1989.

"Java's Agricultural Production, 1775–1875." In *Economic Growth in Indonesia, 1820–1940*, edited by Angus Maddison and Gé Prince, 97–131. Dordrecht/Providence: Foris, 1989.

"Treacherous Cane: The Java Sugar Industry between 1914 and 1940." In *The World Sugar Economy in War and Depression, 1914–1940*, edited by B. Albert and A. Graves, 157–169. London/New York: Methuen, 1988.

"Mengubah Ukuran dan Perubahan Ukuran: Pertumbuhan Pertanian Daerah di P. Java 1815–1875." In *Sejarah Ekonomi Indonesia*, edited by Anne Booth et al., 165–196. Jakarta: LP3ES, 1988.

"Forests and Forestry in Colonial Java, 1677–1942." In *Changing Tropical Forests: historical perspectives on today's challenges in Asia, Australasia and Oceania*, edited by J. Dargaval, K. Dixon and N. Semple, 59–87. Canberra: CRES, 1988.

"Morbidity and Mortality in Java, 1820–1880: Changing Patterns of Disease and Death." In *Death and Disease in Southeast Asia: Explorations in Social, Medical and Demographic History*, edited by N.G. Owen, 48–69. Singapore: Oxford University Press, 1987. [Asian Studies Association of Australia; Southeast Asia Publication Series, 14.]

"Tegen Duizendpoten, Duikboten en Shylock Ruys: Bouwvakstakingen in Nederland tussen de Wereldoorlogen." In *Exercities in ons Verleden: Twaalf Opstellen over de Economische en Sociale Geschiedenis van Nederland en Koloniën, 1800–1950: Aangeboden aan Prof. W.J. Wieringa*, edited by P. Boomgaard et al., 129–152. Assen: Van Gorcum, 1981.

"Methodological Problems in comparing strike Incidence with Long and Short Waves." In *Konjunktur, Krise, Gesellschaft*, edited by D. Petzina and G.v. Roon, 256–259. Stuttgart: Klett Cotta, 1981.

"Bevolkingsgroei en Welvaart op Java (1800–1942)." In *Indonesië Toen en Nu*, edited by R.N.J. Kamerling, 35–52. Amsterdam: Intermediair bibliotheek, 1980.

"Brazilië: Een Reus op Lemen Voeten?" In *Spectrum Jaarboek 1979*, 176–184. Utrecht/Antwerpen: Het Spectrum, 1979.

"Geschiedenis (1873–1975)." In *Encyclopedie van Suriname*, edited by C.F.A. Bruijning and J. Voorhoeve, 240–245. Amsterdam/Brussel: Elsevier, 1977.

"Vrijheid in Afhankelijkheid: 100 jaar Economische Ontwikkeling na het Einde van het Staatstoezicht in Suriname (1873–1973)." In *Suriname: Van Slavernij naar Onafhankelijkheid: De Geschiedenis van Suriname 1674–1974*, 63–82. Amsterdam: OSGN, 1975. [Congres verslag, OSGN wetenschappelijke publicaties nr. 3.]

"Stakingen, Vakbondslidmaatschap en Conjunctuur gedurende het Interbellum in Nederland." In *Lof der Historie, Opstellen over Geschiedenis en Maatschappij*, edited by J. van Herwaarden, 339–373. Rotterdam: Universitaire Pers Rotterdam, 1973.

References

Archival Sources

Archivo General de Indias, Seville (AGI)

Audiencia de Filipinas (AF) 296 Oficiales reales 13 July 1709.
Patronato (PAT) 23–9 Miguel de Loarca [1582].
Patronato (PAT) 23–7 Diego de Artieda, no date.
Patronato (PAT) 25–45 Breve sumario 1593.

Arsip Nasional Republik Indonesia (ANRI)

Makassar Residency archive, 289/2.

Library of Congress (LC)

"Lecture on Forests and Forest Work in the Philippines," Pinchot papers, box 640, file 1902, no. 9, Yale Forest School, 16 March 1903.
"Lecture on Forests and Forest Work in the Philippines First Half," Pinchot papers, box 640, file 1902, no. 10, Yale Forest School, 16 March 1903.
"Lecture on Forests and Forest Work in the Philippines Second Half," Pinchot papers, box 640, file 1902, no. 10, Pinchot, Philippine Islands, Yale Forest School, 16 March 1903.
"Philippine Islands," Pinchot papers, box 640, file: Philippine Islands, 'Philippine Islands'.
"Regino Garcia, Brief Review of the Forestry Service during the Spanish Government from 1863 to 1898," Pinchot Papers, box 586, file: Forestry, Philippine Islands, Bureau of Forestry, General Organization, 1903.
"XII Philippines," Pinchot papers, box 640, file: Philippines, John Lydenberg's account of Gifford Pinchot.

National Archief, The Hague (NA)

VOC collection, 1663/174–186.

National Archives and Records Administration (NARA)

Archives-II (College Park), RG 350, General records of the Bureau of Insular Affairs, box 768, file 11896. John S. Sullivan, Assistant to Chief of Bureau to Coverdale and Colpitts, Consulting Engineers, 31 May 1928.
Archives-II (College Park), RG 350, General records of the Bureau of Insular Affairs, 1914–45, box 272, file 1214–54, no. 3081, "I.X.L. Cattle Co. Inc.," 7 December 1923.
Archives-II (College Park), RG 350, General records of the Bureau of Insular Affairs, 1914–45, box 422, file 2392–149, War Department, Note, 1 June 1927.

Philippine National Archive (PNA)

Ministerio de Ultramar to governor-general of the Philippine Islands, Madrid, 7 August 1877, Corte de maderas, bundle 2.

Published Sources

Acda, Menandro N. 2004. "Economically Important Termites (*Isoptera*) of the Philippines and Their Control." *Sociobiology* 43: 159–168.

Adas, Michael. 1981. "From Avoidance to Confrontation: Peasant Protest in Precolonial and Colonial Southeast Asia." *Comparative Studies in Society and History* 23: 217–247.

Agoncillo, Teodoro. 2001. *The Fateful Years: Japan's Adventure in the Philippines, 1941–45*. Diliman: University of the Philippines Press.

Aguilar, Filomeno V. 1998. *Clash of Spirits: The History of Power and Sugar Planter Hegemony on a Visayan Island*. Honolulu: University of Hawai'i Press.

Ahern, George. 1901. *Compilation of Notes on the Most Important Timber Tree Species of the Philippine Islands*. Manila: Bureau of Forestry.

——. 1913. *Annual Report of the Director of Forestry of the Philippine Islands for the Fiscal Year Ended June 30, 1913*. Manila: Bureau of Printing.

Alarcón, Norma. 1991. *Philippine Architecture during the Pre-Spanish and Spanish Periods*. Manila: Santo Tomas University Press.

Alexander, Paul. 1984. "Women, Labour and Fertility: Population Growth in Nineteenth Century Java." *Mankind* 14 (4): 361–372

Alpers, Svetlana.1983. *The Art of Describing: Dutch Art in the Seventeenth Century*. Chicago: University of Chicago Press.

Amrith, Sunil S. 2011. *Migration and Diaspora in Modern Asia*. Cambridge: Cambridge University Press.

Andaya, Barbara. 1979. *Perak, the Abode of Grace: A Study of an Eighteenth-Century Malay State*. Kuala Lumpur: Oxford University Press.

——. 1993. *To Live as Brothers: Southeast Sumatra in the Seventeenth and Eighteenth Centuries*. Honolulu: University of Hawai'i Press.

Andaya, Leonard. 1975. "The Nature of Kingship in Bone." In *Pre-colonial State Systems in Southeast Asia*, edited by A. Reid and L. Castles, 115–125. Kuala Lumpur: MBRAS.

——. 1981. *The Heritage of Arung Palakka: A History of South Sulawesi (Celebes) in the Seventeenth Century*. The Hague: Martinus Nijhoff.

——. 1991. "Local Trade Networks in Maluku in the 16th, 17th and 18th Centuries." *Cakalele* II (2): 71–96.

Andaya, Barbara and Leonard Andaya. 2001. *A History of Malaysia*. Basingstoke: Palgrave.

Anderson, Benedict. 1983. "Old State, New Society: Indonesia's New Order in Comparative Historical Perspective." *Journal of Asian Studies* 42: 477–496.

Anderson, Kathryn Gay. 2003. *The Open Door: Early Modern Wajorese Statecraft and Diaspora*. PhD thesis, University of Hawai'i.

Antlöv, Hans. 1994. "Village Leaders and the New Order." In *Leadership on Java: Gentle Hints, Authoritarian Rule*, edited by Hans Antlöv and Sven Cederroth, 73–98. Richmond: Curzon.

Appell, George N. 1968. "The Penis Pin at the Peabody Museum, Harvard University." *Journal of the Malaysian Branch Royal Asiatic Society* 41: 203–205.

Ariyasajsiskul, Supaporn. 2004. "The So-Called Tin Monopoly in Ligor: The Limits of VOC Power vis à vis a Southern Thai Trading Polity." *Itinerario* 28 (3): 89–106.

Arsip Nasional Republik Indonesia. 1992. Pameran seabad kearsipan: Menunjang pembangunan bangsa Indonesia. 1892–1992, Jakarta.

Atmodjo, Sukarto K. 1985. "Some Short Notes on Agricultural Data from Ancient Balinese Inscriptions." In *Agrarian History: Papers of the Fourth Indonesian-Dutch History Conference*, edited by Sartono Kartodirdjo, 25–62. Yogyakarta: Gadjah Mada University Press.

Backus, Charles. 1981. *The Nan-Chao Kingdom and T'ang China's Southwestern Frontier*. Cambridge: Cambridge University Press.

Bagley, Robert. 2001. *Ancient Sichuan: Treasures from a Lost Civilization*. Seattle: Seattle Art Museum.

Bakker, Laurens. 2009a. "Adat, Land and Popular Democracy: Dayak Politics in East Kalimantan." *Borneo Research Bulletin* 40: 202–220.

——. 2009b. "Politics and Polls: Elections and Land Conflicts in Paser, Indonesia." *ASEAS* 2: 30–48.

Balfour, Edward. 1871. *Cyclopaedia of India and of Eastern and Southern Asia, Commercial, Industrial and Scientific*. Madras: The Asylum. Second edition.

Bankoff, Greg. 2001. "A Question of Breeding: Zootechny and Colonial Attitudes towards the Tropical Environment in Late Nineteenth Century Philippines." *Journal of Asian Studies* 60: 413–437.

——. 2003. *Cultures of Disaster: Society and Natural Hazard in the Philippines*. London: Routledge.

——. 2007a. "Colonising New Lands: Horses in the Philippines." In *Breeds of Empire: The 'Invention' of the Horse in Southeast Asia and Southern Africa, 1500–1950*, edited by Greg Bankoff and Sandra Swart, 85–103. Copenhagen: NIAS Press.

——. 2007b. "Adapting to a New Environment: The Philippine Horse." In *Breeds of Empire: The 'Invention' of the Horse in Southeast Asia and Southern Africa, 1500–1950*, edited by Greg Bankoff and Sandra Swart, 105–121. Copenhagen: NIAS Press.

——. 2007c. "One Island Too Many: Reappraising the Extent of Deforestation in the Philippines Prior to 1946." *Journal of Historical Geography* 33: 314–334.

——. 2012. "A Tale of Two Cities: The Pyro-seismic Morphology of Nineteenth Century Manila." In *Flammable Cities: Urban Fire and the Making of the Modern World*, edited by Greg Bankoff, Uwe Luebken and Jordan Sand, 170–189. Madison: University of Wisconsin Press.

Barlow, Colin.1990. *Development in Eastern Indonesia: The Case of Nusa Tenggara Timur.* Canberra: Australian Government Publishing Service.

Barrera-Osorio, Antonio. 2006. *Experiencing Nature: The Spanish American Empire in Early Modern Europe.* Austin: University of Texas.

Barwegen, Martine. 2005. *Gouden Hoorns: De Geschiedenis van de Veehouderij op Java, 1850–2000*. PhD Thesis, Wageningen University.

Bateson, Gregory. 1972. *Steps to an Ecology of Mind: Collected Essays in Anthropology, Psychiatry, Evolution, and Epistemology*. Chicago: University of Chicago Press.

Bätschmann, Oskar. 2008. *Giovanni Bellini*. London: Reaktion Books.

Bayle, Constantino. 1949. "Informe Sobre los Naturales de la Diócesis de Cebu." *Archivo Missionalia Hispánica* 6: 389–398.

Bell, Charles. 1998. *The People of Tibet*. Delhi: Book Faith. Reprint of 1928.

Bellina, Bérénice, Elisabeth A. Bacus, Thomas Oliver Pryce and Jan Wisseman Christie. eds. 2010. *50 Years of Archaeology in Southeast Asia: Essays in Honour of Ian Glover.* Bangkok: River Books.

Bellwood, Peter. 1985. *Prehistory of the Indo-Malaysian archipelago*. Sydney: Academic Press.

Bemmelen, R.W. van. 1956. "The Influence of Geologic Events on Human History (An Example from Central Java)." *Verhandelingen van het Koninklijk Nederlandsch Geologisch-Mijnbouwkundig Genootschap, Geologische Serie* 16: 19–36.

Benda, Harry J. 1964. "Democracy in Indonesia: The Decline of Constitutional Democracy in Indonesia. By Herbert Feith." *The Journal of Asian Studies* 23-3: 449–456. Review article.

——. 1966. "The Pattern of Administrative Reforms in the Closing Years of Dutch Rule in Indonesia." *Journal of Asian Studies* 25: 589–605.

Benson, Sarah. 2011. "European Wonders at the Court of Siam." In *Collecting Across Cultures: Material Exchanges in the Early Modern Atlantic World*, edited by Daniela Bleichmar and Peter C. Mancall, 155–177. Philadelphia: University of Pennsylvania Press.

Berger, John. 1974. *The Look of Things: Essays.* New York: Viking Press.

Berry, Brian. 1964. "Cities as Systems within Systems of Cities." *Papers in Regional Science* 13: 147–164.

Biografisch portal van Nederland, www.biografischportaal.nl.

Blair, Emma H. and James A. Robertson. eds. 1903–1907. *The Philippine Islands, 1493–1803*. 55 vols. Cleveland, OH: The Arthur H. Clark Co.

Blench, Roger. 2010. "The Spread of the Horse into Southeast Asia: Evidence from Vernacular Names." Unpublished paper, Cambridge, Kay Williamson Educational Foundation.

Bleichmar, Daniela and Peter C. Mancall. eds. 2011. *Collecting Across Cultures: Material Exchanges in the Early Modern Atlantic World.* Philadelphia: University of Pennsylvania Press.

Bloch, Marc. 1964. *Feudal Society.* 2 vols. Chicago: University of Chicago Press.

Blussé, Leonard. 1986. *Strange Company: Chinese Settlers, Mestizo Women, and the Dutch in VOC Batavia.* Dordrecht-Holland: Foris Publications.

——. 1991. "In Praise of Commodities: An Essay on the Crosscultural Trade in Edible Bird's-Nests." In *Emporia, Commodities, and Entrepreneurs in Asian Maritime Trade, c. 1400–1750*, edited by Roderich Ptak and Dietmar Rothermund, 317–335. Stuttgart: Steiner Verlag.

——. 1997. *Bitters Bruid: Een Koloniaal Huwelijksdrama in de Gouden Eeuw.* Amsterdam: Balans.

——. 1999. "Chinese Century: The Eighteenth Century in the China Sea Region." *Archipel* 58: 107–129.

——. 2011. "Junks to Java: Chinese Shipping to the Nanyang in the Second Half of the Eighteenth Century." In *Chinese Circulations: Capital, Commodities and Networks in Southeast Asia*, edited by Eric Tagliacozzo and Chang Wen-chin, 221–258. Durham: Duke University Press.

Bøckman, Harald. 1991. "Yunnan Trade in Han Times: Transit, Tribute and Trivia." In *Asian Trade Routes*, edited by Karl R. Haellquist, 174–180. London: Curzon.

Boechari, M. 1985. "Prasasti Tihang." Unpublished typescript.

——. 1986. *Prasasti Koleksi Museum Nasional.* Jakarta: Museum Nasional.

Boeke, J.H. 1940. *Indische Economie.* 2 vols. Haarlem: H.D. Tjeenk Willink.

——. 1953. *Economics and Economic Policy of Dual Societies as Exemplified by Indonesia.* Haarlem: H.D. Tjeenk Willink.

Boers de Jong, Bernice. 1997. "Paardenfokkerij op Sumbawa, 1500–1930." *Spiegel Historiael, Magazine voor Geschiedenis en Archeologie* 32 (10–11): 438–443.

——. 2007. "The 'Arab' of the Indonesian Archipelago: Famed Horse Breeds of Sumbawa." In *Breeds of Empire: The 'Invention' of the Horse in Southeast Asia and Southern Africa 1500–1950*, edited by Greg Bankoff and Sandra Swart, 51–64. Copenhagen: NIAS Press.

Bondevik, Stein. 2008. "Earth Science: The Sands of Tsunami Time." *Nature* 455: 1183–1184.

Boomgaard, Peter. 1988. "Forests and Forestry in Colonial Java, 1677–1942." In *Changing Tropical Forests: Historical Perspectives on Today's Challenges in Asia, Australasia and Oceania*, edited by John Dargavel, Kay Dixon and Noel Semple, 58–87. Canberra: Centre for Resource and Environmental Studies, Australian National University.

——. 1989a. *Children of the Colonial State: Population Growth and Economic Development in Java, 1795–1880.* Amsterdam: Free University Press.

——. 1989b. "The Javanese Rice Economy 800–1800." In *Economic and Demographic Development in Rice Producing Societies: Some Aspects of East Asian Economic History, 1500–1900*, edited by Peter Boomgaard, Akira Hayami and Yoshihiro Tsubouchi, 317–344. Tokyo: [n.p.].

——. 1991. "The Javanese Village as a Cheshire Cat: The Java Debate against a European and Latin American Background." *The Journal of Peasant Studies* 18 (2): 288–304.

——. 1993. "Economic Growth in Indonesia, 500–1990." In *Explaining Economic Growth: Essays in Honour of Angus Maddison*, edited by Adam Szirmai, Bart van Ark and Dirk Pilat, 195–216. Amsterdam: Elsevier Science.

——. 1996. *Forests and Forestry 1823–1941*. Amsterdam: KIT.

——. 1998a. "The VOC Trade in Forest Products in the 17th Century." In *Nature and the Orient; The Environmental History of South and Southeast Asia*, edited by R.H. Grove, V. Damodaran and S. Sangwan, 375–395. Delhi: Oxford University Press.

——. 1998b. "Environmental Impact of the European Presence in Southeast Asia, 17th–19th Centuries." *Illes i Imperis* 1: 21–35.

——. 1999. "Maize and Tobacco in Upland Indonesia, 1600–1940." In *Transforming the Indonesian Uplands; Marginality, Power and Production*, edited by Tania Murray Li, 45–78. Amsterdam: Harwood Academic Publishers.

——. 2001a. *Frontiers of Fear; Tigers and People in the Malay world, 1600–1950*. New Haven: Yale University Press.

——. 2001b. "Crisis Mortality in Seventeenth-Century Indonesia." In *Asian Population History*, edited by Ts'ui-jung Liu et al., 191–220. Oxford: Oxford University Press.

——. 2002. "From Subsistence Crises to Business Cycle Depressions, Indonesia 1800–1940." *Itinerario* 26 (3/4): 35–49.

——. 2003a. "Bridewealth and Birth Control: Low Fertility in the Indonesian Archipelago, 1500–1900." *Population and Development Review* 29 (2): 197–214.

——. 2003b. "Human Capital, Slavery and Low Rates of Economic and Population Growth in Indonesia, 1600–1910." *Slavery and Abolition* 24 (2): 83–96.

——. 2003c. "Smallpox, Vaccination, and the 'Pax Neerlandica'; Indonesia, 1550–1930." *Bijdragen tot de Taal-, Land- en Volkenkunde* 159: 590–617.

——. 2004. "Horses, Horse Trading and Royal Courts in Indonesian History, 1500–1900." In *Smallholders and Stockbreeders: Histories of Foodcrop and Livestock Farming in Southeast Asia*, edited by Peter Boomgaard and David Henley, 211–232. Leiden: KITLV Press.

——. 2005a. "Products and People of the Sea in and around the Indonesian Archipelago, 900–1900." In *Muddied Waters; Historical and Contemporary Perspectives on the Management of Forests and Fisheries in Island Southeast Asia*, edited by Peter Boomgaard, David Henley, and Manon Osseweijer, 97–120. Leiden: KITLV Press.

——. 2005b. "The Long Goodbye? Trends in Forest Exploitation in the Indonesian Archipelago, 1600–2000." In *Muddied Waters; Historical and Contemporary Perspectives on the Management of Forests and Fisheries in Island Southeast Asia*, edited by Peter Boomgaard, David Henley, and Manon Osseweijer, 211–235. Leiden: KITLV Press.

——. 2007a. "Horse Breeding, Long-Distance Horse Trading, and Royal Courts in Indonesian History, 1500–1900." In *Breeds of Empire: The 'Invention' of the Horse in Southeast Asia and Southern Africa, 1500–1950*, edited by Greg Bankoff and Sandra Swart, 33–49. Copenhagen: NIAS Press.

——. 2007b. *Southeast Asia: An Environmental History*. Santa Barbara, California: ABC-CLIO.

——. ed. 2007c. *A World of Water: Rain, Rivers, and Seas in Southeast Asian Histories*. Leiden: KITLV Press.

——. 2008a. "Technologies of a Trading Empire: Dutch Introduction of Water- and Windmills in Early-Modern Asia, 1650s–1800." *History and Technology* 24 (1): 41–59.

——. 2008b. "Human Capital, Slavery and Low Rates of Economic and Population Growth in Indonesia, 1600–1910." *Slavery and Abolition* 24 (2): 83–96.

——. 2008c. "Early Globalization: Cowries as Currency, 600 BCE–1900." In *Linking Destinies: Trade, Towns and Kin in Asian History*, edited by Peter Boomgaard, Dick Kooiman and Henk Schulte Nordholt, 13–28. Leiden: KITLV Press.

——. 2011. "Land Rights and the Environment in the Indonesian Archipelago, 800–1950." *Journal of the Economic and Social History of the Orient* 54 (4): 478–496.

Boomgaard, Peter and Janneke van Dijk. 2001. *Het Indië Boek*. Zwolle: Waanders.

Bosch, F.D.K. 1961. *Selected Studies in Indonesian Archaeology*. The Hague: Martinus Nijhoff.

Bosma, Ulbe and Remco Raben. 2008. *Being "Dutch" in the Indies: A history of creolization and empire, 1500–1920*, Singapore: National University of Singapore Press.

Boxer, C.R. 1967. *Francisco Vieira De Figueiredo: A Portuguese Merchant-Adventurer in South East Asia, 1624–1667*. Vol. 52. 's-Gravenhage: Nijhoff.

Brandes, J.L.A. 1899. *Notulen van het Bataviaasch Genootschap van Kunsten en Wetenschappen*. Notes 3, Batavia.

——. 1913. *Oud-Javaansche Oorkonden: Nagelaten Transcripties*. Batavia: Albrecht, 's-Gravenhage: Nijhoff.

Braudel, Fernand. 1949. *La Méditerranée et le Monde Méditerranéen a l'Époque de Philippe II*. 3 vols. Paris: Armand Colin.

——. 1958. "Histoire et Sciences Sociales: La Longue Durée." *Annales: Economies, Societies, Civilisations* 4 (1): 725–753.

——. 1960. "History and the Social Sciences: The Long Duration." *American Behavioural Scientist* 3: 3–13.

——. 1972–1973. *The Mediterranean and the Mediterranean World in the Age of Philip II*. 2 vols. London: Collins.

——. 1980. *On History*. Translated by Sarah Matthews. Chicago: University of Chicago Press.

Breman, Jan. 1980. *The Village on Java and the Early Colonial State.* Rotterdam: CASP, Faculty of Social Sciences, Erasmus University.

Broad, Robin and John Cavanagh. 1993. *Plundering Paradise: The Struggle for the Environment in the Philippines.* Berkeley: University of California Press.

Brown, Donald E., James W. Edwards and Ruth P. Moore. 1988. *The Penis Inserts of Southeast Asia: An Annotated Bibliography with an Overview and Comparative Perspectives.* Berkeley: Center for Southeast Asia Studies, University of California.

Brown, Roxanna M. ed. 1989. *Guangdong Ceramics from Butuan and Other Philippine Sites.* Manila: Oriental Ceramic Society of the Philippines and Oxford University Press.

Bruijn, Max de and Remco Raben. eds. 2004. *The World of Jan Brandes, 1743–1808: Drawings of a Dutch Traveller in Batavia, Ceylon and Southern Africa.* Zwolle: Waanders, Amsterdam: Rijksmuseum.

Bruinessen, Martin van. 1991. "The Tariqa Khalwatiyya in South Celebes." In *Excursies in Celebes*, edited by Harry A. Poeze and Pim Schoorl, 251–271. Leiden: KITLV.

Bryant, E., G. Walsh and D. Abbott. 2007. "Cosmogenic Mega-tsunami in the Australia Region: Are They Supported by Aboriginal and Maori Legends?" University of Wollongong. Accessed June 2012. http://ro.uow.edu.au/cgi/viewcontent.cgi?article=1035&context=scipapers.

Bryant, Ralph. 1907. "Some Philippine Forest Problems." *Proceedings of the Society of American Foresters* 2 (1): 3–19.

Buckley, Brendan M. et al. 2010. "Climate as a Contributing Factor in the Demise of Angkor, Cambodia." *Proceedings of the National Academy of Sciences of the United States of America* 107 (15): 6748–6752.

Buehler, Michael. 2007. "Rise of the Clans." *Inside Indonesia* 90.

——. 2009. "The Rising Importance of Personal Networks in Indonesian Local Politics. An Analysis of District Government Head Elections in South Sulawesi in 2005." In *Deepening Democracy in Indonesia? Direct Elections for Local Leaders (Pilkada)*, edited by Maribeth Erb and Priyambudi Sulistiyanto, 101–124. Singapore: ISEAS.

Bulbeck, David. 1994. *Ecological Parameters of Settlement Patterns and Hierarchy in the Pre-colonial Makassar Kingdom.* Perth: ASAA Biennial Conference.

Bulbeck, David and Ian Caldwell. 2008. "Oryza Sativa and the Origins of Kingdoms in South Sulawesi, Indonesia: Evidence from Rice Husk Phytoliths." *Journal of Indonesia and the Malay World* 36 (104): 1–20.

Bulbeck, F.D. 1992. *A Tale of Two Kingdoms: The Historical Archaeology of Gowa and Tallok, South Sulawesi, Indonesia.* PhD thesis, Australian National University.

Bulbeck, F. David. 1996. "The Politics of Marriage and the Marriage of Polities in Gowa, South Sulawesi, During the 16th and 17th Centuries." In *Origins, Ancestry and Alliance*, edited by James J. Fox and Clifford Sather, 280–315. Canberra: Anthropology, RSPS Australian National University.

Burke, Peter. 2001. *Eyewitnessing: The Uses of Images as Historical Evidence*. London: Reaktion Books.

Burkill, I.H. 1966. *A Dictionary of the Economic Products of the Malay Peninsula*. 2 vols. Kuala Lumpur: Ministry of Agriculture and Co-operatives. First published in 1935.

Burzynski, Joseph. 2002. "The Timber Trade and the Growth of Manila, 1864–1881." *Philippine Studies* 50 (2): 168–192.

Cabaton, A. 1911. *Java, Sumatra and the Other Islands of the Dutch East Indies*. London: T. Fisher Unwin.

Caldwell, Ian. 1991. "The Myth of the Exemplary Centre; Shelly Errington's Meaning and Power in a Southeast Asian Realm." *Journal of Southeast Asian Studies* 22 (1): 109–188.

——. 1995. "Power, State and Society among the Pre-Islamic Bugis." *Bijdragen tot de Taal-, Land- en Volkenkunde* 151 (3): 394–421.

Callwell, C.E. 1990. *Small Wars, a Tactical Textbook for Imperial Soldiers*. London: Greenhill Books. Reprint of 3rd edition, 1906.

Cameron, John. 1865. *Our Tropical Possessions in Malayan India*. London: Smith.

Campany, Robert Ford. 1996. *Strange Writing: Anomaly Accounts in Early Medieval China*. Albany: State University of New York.

Campbell, Mary B. 2004. *Wonder & Science: Imagining Worlds in Early Modern Europe*. Ithaca: Cornell University Press.

Campen, Jan van and Ebeltje Hartkamp-Jonxis. 2011. *Asian Splendour: Company Art in the Rijksmuseum*. Zutphen: Walburg Pers.

Campo, J.N.F.M.à. 2002. *Engines of Empire: Steamshipping and State Formation in Colonial Indonesia*. Hilversum: Verloren.

Carey, P. 1986. "Waiting for the 'Just King': The Agrarian World of South-Central Java from Giyanti (1755) to the Java War (1825–1830)." *Modern Asian Studies* 20 (1): 59–137.

Carletti, Francesco. 1965. *My Voyage Around the World*. Translated by Herbert Weinstock. London: Methuen and Co. Ltd.

Casparis, J.G. de. 1956. *Prasasti Indonesia II. Selected Inscriptions from the 7th to the 9th Century* A.D. Bandung: Masa Baru.

——. 1958. "Short Inscriptions from Tjandi Plaosan-Lor." *Berita Dinas Purbakala/ Bulletin of the Archaeological Service of the Republic of Indonesia* 4: 1–36.

——. 1988. "Where Was Pu Siṇḍok's Capital Situated?" *Studies in South and Southeast Asian Archaeology* 2: 39–52. Leiden: Koentji Press.

Cavendish, Thomas. 1890. "The Admirable and Prosperous Voyage of the Worshipful Master Written by Master Francis Pretty 1586–1588." In *The Principal Navigations, Voyages, Traffiques and Discoveries of the English Nation, Collected by Richard Hakluyt*, edited by E. Goldsmid, 5–84. London: E and G. Goldsmid.

CDIU. 1885–1932. *Colección de Documentos Inéditos Relativos al Descubrimiento, Conquista y Organización de las Antiguas Posesiones Españoles de Ultramar*. 25 vols. 2nd series. Madrid: Real Academia de la Historia.

Census of the Philippine Islands. 1905. 4 vols. Washington, DC: US Bureau of the Census.

Chabot, Hendrik. 1950. *Verwantschap, Stand en Sexe in Zuid-Celebes*. Djakarta: Balai Perguruan Tinggi Nasional.

Chabot, H. Th. Kinship. 1996. *Status and Gender in South Celebes*. Leiden: KITLV.

Chaiklin, Martha. 2003. *Cultural Commerce and Dutch Commercial Culture: The Influence of European Material Culture on Japan, 1700–1850*, 5. Leiden: Research School CNWS. Studies in Overseas History.

Chakravarti, Ranabir. 1999. "Early Medieval Bengal and the Trade in Horses: A Note." *Journal of the Economic and Social History of the Orient* 42 (2): 194–211.

——. 2009. "Equestrian Demand and Dealers: The Early Indian Scenario, up to c. 1300." In *Pferde in Asien: Geschichte, Handel und Kultur*, edited by Bert Fragner, Ralph Kauz, Roderich Ptak and Angela Schottenhammer, 145–159. Vienna: Verlag der Österreichischen Akademie der Wissenschaften.

Chambert-Loir, Henri. 2009. "Le Dit du Sultan de Barus." In *Histoire de Barus III: Regards sur une Place Marchande de l'Océan Indien (XXIIe-milieu du XVIIe s.), Cahiers d'Archipel 38*, edited by Daniel Perret and Heddy Surachman, 507–528. Paris: Association Archipel, École française d'Extrême-Orient.

Chao, Ju Kua [Zhao Rugua]. 1911. *Chau Ju-kua: His Work on the Chinese and Arab Trade in the Twelfth and Thirteenth Centuries, Entitled Chu-Fan-Chï*. Translated and annotated by Friedrich Hirth and W.W. Rockhill. St. Petersburg: Imperial Academy of Sciences.

Chandola, Khemanand. 1987. *Across the Himalayas through the Ages: A Study of Relations between Central Himalayas and Western Tibet*. New Delhi: Patriot.

Chandra, L. 1997. "Śaiva Version of Sang Hyang Kamahāyānikan." In *Cultural Horizons of India*, edited by L. Chandra, 7–101. Vol. 5. New Delhi: International Academy of Indian Culture.

Charney, Michael. 2004. *Southeast Asian Warfare, 1300–1900*. Leiden: Brill.

Chase-Dunn, Christopher K. and Thomas D. Hall. 1997. *Rise and Demise: Comparing World-Systems*. Boulder, Colo: Westview Press.

Chen Chingho. 1977. "Mac Thien Tu and Phraya Taksin: A Survey on Their Political Stand, Conflicts, and Background." In *Proceedings of the Seventh IAHA Conference*, 1534–1575. Bangkok: Chulalongkorn University.

Chou, Ta-kuan [Zhou Daguan]. 1967. *Notes on the Customs of Cambodia*. Bangkok: Social Science Association Press.

Chirino, Pedro de. [1604] 1969. *Relación de las Islas Filipinas*. Manila: Historical Conservation Society.

Christie, Jan Wisseman. 1991. "States without Cities: Demographic Trends in Early Java." *Indonesia* 52: 23–40.

——. 1992. "Water from the Ancestors: Irrigation in Early Java and Bali." In *The Gift of Water: Water Management, Cosmology and the State in South East Asia*, edited by J. Rigg, 7–25. London: School of Oriental and African Studies.

——. 1993. "Trade and value in pre-Majapahit Java." *Indonesia Circle* 59/60: 3–17.

——. 1996. "Money and Its Uses in the Javanese States of the Ninth to Fifteenth Centuries A.D." *Journal of the Economic and Social History of the Orient* 39 (3): 243–286.

——. 1998. "Javanese Markets and the Asian Sea Trade Boom of the Tenth to Thirteenth Centuries A.D." *Journal of the Economic and Social History of the Orient* 41 (3): 344–381.

——. 1999. "Asian Sea Trade between the Tenth and Thirteenth Centuries and Its Impact on the States of Java and Bali." In *Archaeology of Seafaring: The Indian Ocean in the Ancient Period*, edited by H.P. Ray, 221–270. Delhi: Pragati Publications for the Indian Council of Historical Research.

——. 2001. "Revisiting Early Mataram." In *Fruits of Inspiration: Studies in Honour of Prof. J.G. de Casparis*, edited by M.J. Klokke, and K.R. van Kooij, 25–55. Groningen: Egbert Forsten.

——. 2003. "Sobhāmṛta." Unpublished manuscript.

Christensen, Norman L. 1989. "Landscape History and Ecological Change." *Journal of Forest History* 33: 116–125.

Clarence-Smith, William G. 2002. "Horse Trading: The Economic Role of Arabs in the Lesser Sunda Islands, c. 1800 to c. 1940." In *Transcending Borders: Arabs, Politics, Trade and Islam in Southeast Asia*, edited by Huub de Jonge and Nico Kaptein, 143–162. Leiden: KITLV Press.

——. 2004a. "Cape to Siberia: The Indian Ocean and China Sea Trade in Equids." In *Maritime Empires: British Imperial Maritime Trade in the Nineteenth Century*, edited by David Killingray, Margarette Lincoln and Nigel Rigby, 48–67. Woodbridge: Boydell and Brewer.

——. 2004b. "Horse Breeding in Mainland Southeast Asia and its Borderlands." In *Smallholders and Stockbreeders: Histories of Foodcrop and Livestock Farming in Southeast Asia*, edited by Peter Boomgaard and David Henley, 189–210. Leiden: KITLV Press.

——. 2004c. "Elephants, Horses, and the Coming of Islam to Northern Sumatra." *Indonesia and the Malay World* 32 (93): 271–284.

Clodfelder, Micheal. 1992. *Warfare and Armed Conflicts: A Statistical Reference to Casualty and other Figures, 1618–1991*. Jefferson, N.C.: McFarland.

Coates, Austin. 1994. *China Races*. Hong Kong: Oxford University Press. Second edition.

Cohen Stuart, Abraham Benjamin. 1875. *Kawi Oorkonden in Facsimile: Met Inleiding en Transcriptie*. Leiden: E.J. Brill.

Colín, Pedro. 1900–1902. *Labor Evangélica*. 3 vols. Barcelona: Heinrich y Compañía. Edited by Pablo Pastells.

Collins, Thomas. 1901. "Testimony of Thomas Collins to the Philippine Commission, 12 May 1899." In *Report of the Philippine Commission to the President*. Washington: Government Printing Office.

Colombijn, Freek. 1994. *Patches of Padang: The History of an Indonesian Town in the Twentieth Century and the Use of Urban Space*. PhD thesis, Leiden University.

Combes, Francisco. 1897. *Historia de Mindanao y Jolo*. Madrid: Pablo Pastells and W.E. Retana.

Conklin, Harold. 1957. *Hanunóo Agriculture: A Report on an Integral System of Shifting Cultivation in the Philippines.* Rome: Food and Agriculture Organization.

Cook, Harold J. 2001. "Time's Bodies: Crafting the Preparation and Preservation of Naturalia." In *Merchants and Marvels: Commerce, Science and Art in Early Modern Europe*, edited by Pamela Smith and Paula Findlen, 223–247. New York, NY: Routledge.

——. 2005. "Global Economies and Local Knowledge in the East Indies: Jacobus Bontius Learns the Facts of Nature." In *Colonial Botany: Science, Commerce, and Politics in the Early Modern World*, edited by Claudia Swan and Londa Schiebinger, 100–118. Philadelphia: Penn State University Press.

——. 2007. *Matters of Exchange: Commerce, Medicine, and Science in the Dutch Golden Age.* New Haven: Yale University Press.

Cook, Harold J. and Timothy D. Walker. eds. 2013. "Circulation of Medicine in the Early Modern Atlantic World." *Social History of Medicine* (May): 1–15. Special issue.

Cooke, Nola and Li Tana. eds. 2004. *Water Frontier: Commerce and the Chinese in the Lower Mekong Region, 1750–1880.* Singapore: University Press, Oxford: Rowman and Littlefield.

Cooke, Nola, Li Tana and James A. Anderson. eds. 2011. *The Tongking Gulf through History*. Philadelphia: University of Pennsylvania Press.

Coronel, Hernando de los Rios. 1903–1909. "Memorial y Relacion para su Magestad." In *The Philippine Islands, 1493–1803; Vol. 19*, edited by Emma Blair and Alexander Robertson, 183–300. Cleveland: A.H. Clark Co. First published 1621.

Corpuz, Onofre. 1989. *The Roots of the Filipino Nation*. 2 vols. Vol.1. Quezon City: Aklahi Foundation.

Coster-Wijsman, L.M. 1953. "Illustrations in a Javanese Manuscript." *Bijdragen tot de Taal-, Land- en Volkenkunde* 109 (2): 153–163.

Courtellemont, Gervais. 1904. *Voyage au Yunnan*. Paris: Plon.

Crawfurd, John. 1971. *A Descriptive Dictionary of the Indian Islands and Adjacent Countries*. Kuala Lumpur: Oxford University Press.

Crouch, Harold. 1979. "Patrimonialism and Military Rule in Indonesia." *World Politics* 31: 571–587.

——. 2010. *Political Reform in Indonesia after Soeharto*. Singapore: ISEAS.

Cummings, William. 2001. "Scripting Islamization: Arabic Texts in Early Modern Makassar." *Ethnohistory* 48 (4): 559–586.

——. 2002. *Making Blood White: Historical Transformations in Early Modern Makassar.* Honolulu: University of Hawai'i Press.

——. 2007. "Islam, Empire and Makassarese Historiography in the Reign of Sultan Ala'uddin (1593–1639)." *Journal of Southeast Asian Studies* 38 (2): 197–214.

Cushman, Jennifer. 1993. *Fields from the Sea: Chinese Junk Trade with Siam during the Late Eighteenth and Early Nineteenth Centuries*. Ithaca, NY: Southeast Asia Program, Cornell University.

Cushner, Nicholas. 1971. *Spain in the Philippines: From Conquest to Revolution*. Quezon City: Ateneo de Manila University Press.

DaCosta Kaufmann, Thomas and Michael North. eds. 2014. *Mediating Netherlandish Art and Material Culture in Asia*. Chicago: University of Chicago Press.

Damais, L.-C. 1951. "La Date des Inscriptions en Ère de Sañjaya." *Bulletin de l'École Française d'Extrême-Orient* 45 (1): 42–63.

——. 1952. "Études d'Épigraphie Indonésienne." *Bulletin de l'École Française d'Extrême-Orient* 45 (1): 1–290.

——. 1970. *Répertoire onomastique de l'épigraphie javanaise*. Paris: École Française d'Extrême-Orient.

Dammerman, K.W. 1929. *The Agricultural Zoology of the Malay Archipelago: The Animals Injurious and Beneficial to Agriculture, Horticulture and Forestry in the Malay Peninsular, the Dutch East Indies and the Philippines*. Amsterdam: J.H. de Bussy.

Daniélou, A. 1964. *Hindu Polytheism*. New York: Pantheon Books.

Data Book. 2000. *Data Book on Asian Natural Disasters in the 20th Century*. Kobe: Asian Disaster Reduction Center.

D'Arrigo, Rosanne, et al. 2011. "Three Centuries of Myanmar Monsoon Climate Variability Inferred from Teak Tree Rings." *Geophysical Research Letters* 38: L24705.

Daston, Lorraine and Katherine Park. 1998. *Wonders and the Order of Nature, 1150–1750*. New York: Zone Books.

Davies, H.R. 1909. *Yün-Nan, the Link between India and the Yangtze*. Cambridge: Cambridge University Press.

Day, Tony. 2002. *Fluid Iron: State formation in Southeast Asia*. Honolulu: University of Hawai'i Press.

De Bevoise, Ken. 1995. *Agents of the Apocalypse: Epidemic Disease in the Colonial Philippines*. Princeton: Princeton University Press.

Degroot, V. de. 2008. "Ancient Bathing Places of Central Java: A Short Survey." *Aziatische Kunst* 38 (4): 62–68.

Dent, Anthony A. 1972. *Donkey: The Story of the Ass, from East to West*. London: George G. Harrap.

Dewey, Alice. 1962. *Peasant Marketing in Java*. Glencoe: Free Press.

Diaz-Trechuelo, Maria Lourdes. 1963. "The Economic Development of the Philippines in the Second Half of the Eighteenth Century." *Philippine Studies* 11 (2): 195–231.

Dick, Howard. 2002. "State, Nation-State and National Economy." In *The Emergence of a National Economy: An Economic History of Indonesia 1800–2000*, edited by Howard Dick, Vincent H. Houben, J. Thomas Lindblad and Thee Kian Wie, 9–34. St Leonards: Allen & Unwin, Leiden: KITLV Press.

Dick, Howard and Peter Rimmer. 2003. *Cities, Transport and Communications: The Integration of Southeast Asia since 1850*. Basingstoke: Palgrave Macmillan.

Dijk, Kees van. 2009. "Religion and the Undermining of British Rule in South and Southeast Asia during the Great War." In *Islamic Connections: Muslim Societies in*

South and Southeast Asia, edited by R. Michael Feener and Terenjit Sevea, 109–133. Singapore: ISEAS.

Dijk, Janneke van, Jaap de Jonge and Nico de Klerk. eds. 2010. *J.C. Lamster, een Vroege Filmer in Nederlands-Indië*. Amsterdam: KIT Publishers.

Dirlik, Arif. 1997. "Critical Reflections on Chinese Capitalism as Paradigm." *Identities: Global Studies in Culture and Power* 3: 303–330.

Djoko Dwiyanto. 1986. "Pengamatan Terhadap Data Kesejarahan dari Prasasti Wanua Tengah III tahun 908 M." *Pertemuan Ilmiah Arkeologi* IV (IIa): 92–110. Jakarta: Pusat Penelitian Arkeologi Nasional.

Dobbin, Christine. 1983. *Islamic Revivalism in a Changing Peasant Economy*. London: Curzon Press.

Doeppers, Daniel F. 1968. "Hispanic Influences on Demographic Patterns in the Central Plain of Luzon, 1565–1780." *University of Manila Journal of East Asiatic Studies* 12: 11–96.

Donkin, R.A. 1998. *Beyond Price: Pearls and Pearl-Fishing: Origins to the Age of Discoveries*. Philadelphia: American Philosophical Society.

Douglas, Mary. 1966. "Population Control in Primitive Groups." *British Journal of Sociology* 17 (13): 263–273.

Drakard, Jane. 1990. *A Malay Frontier: Unity and Duality in a Sumatran Kingdom*. Ithaca: Cornell University Southeast Asia Project.

Drieënhuizen, Caroline. 2012. *Koloniale Collecties, Nederlands Aanzien: De Europese Elite van Nederlands-Indië belicht door haar Verzamelingen, 1811–1957*. PhD thesis, University of Amsterdam.

Druce, Stephen C. 2009. *The Lands West of the Lakes: A History of the Ajattappareng Kingdoms of South Sulawesi, 1200 to 1600 CE*. Leiden: KITLV Press.

Edgerton, Ronald. 1982. "Frontier Society on the Bukidnon Plateau." In *Philippine Social History: Global Trade and Local Transformations*, edited by Alfred McCoy and Ed de Jesus, 361–390. Quezon City: Ateneo de Manila University Press.

Elliott, Charles B. 1968. *The Philippines to the End of the Commission Government*. New York: Greenwood Press.

Enenkel, Karl A.E. and Paul J. Smith. eds. 2007. *Early Modern Zoology: The Construction of Animals in Science, Literature and the Visual Arts*. Leiden: Brill.

Errington, Shelley. 1989. *Meaning and Power in a Southeast Asian Realm*. Princeton: Princeton University Press.

Evans, Peter, Dietrich Rueschemeyer and Theda Skocpol. eds. 1985. *Bringing the State Back in*. Cambridge: Cambridge University Press.

Fathurahman, Oman. 2012. "Sifat dan Kekhasan Koleksi Naskah Dayah Tanoh Abee." http://manassa.org/pengantar%20oman.pdf. Accessed July 2012.

Feith, Herbert. 1962. *The Decline of Constitutional Democracy in Indonesia*. Ithaca: Cornell University Press.

Fernández-Armesto, Felipe. 2002. "Epilogue: What Is History *Now?*" In *What Is History Now?* edited by David Cannadine, 148–162. New York: Palgrave Macmillan.

Fischer, Arthur. 1923. *Annual Report of the Director of Forestry of the Philippine Islands for the Fiscal Year Ended December 31, 1922*. Manila: Bureau of Printing.

——. 1925. *Annual Report of the Director of Forestry of the Philippine Islands for the Fiscal Year Ended December 31, 1924*. Manila: Bureau of Printing.

——. 1927. *Annual Report of the Director of Forestry of the Philippine Islands for the Fiscal Year Ended December 31, 1926*. Manila: Bureau of Printing.

——. 1928. *Annual Report of the Director of Forestry of the Philippine Islands for the Fiscal Year Ended December 31, 1927*. Manila: Bureau of Printing.

——. 1929. *Annual Report of the Director of Forestry of the Philippine Islands for the Fiscal Year Ended December 31, 1928*. Manila: Bureau of Printing.

——. 1937. *Annual Report of the Director of Forestry of the Philippine Islands, 1936*. Manila: Bureau of Printing.

Flecker, Michael. 2004–2005. "Treasure from the Java Sea: The 10th Century Intan Shipwreck." *Heritage Asia Magazine* 2 (2): 1–4.

Flores, Jorge. 2007. "Distant Wonders: The Strange and the Marvelous between Mughal India and Habsburg Iberia in the Early Seventeenth Century." *Comparative Studies in Society and History* 49 (3): 553–581.

Forbes, Andrew D.W. 1987. "The 'Cin-Hô' (Yunnanese Chinese) Caravan Trade with Thailand in the Late Nineteenth and Early Twentieth Centuries." *Journal of Asian History* 21 (1): 1–47.

Forbes, Andrew D.W. and David Henley. 1997. *The Haw, Traders of the Golden Triangle*. Chiang Mai: Teak House.

Forbes, W. Cameron. 1976. *The Philippine Islands*. New York: Klaus Reprint Co. First published 1945.

Foreman, John. 1890. *The Philippine Islands*. London: Sampson Low, Marston, Searle & Rivington Ltd.

——. 1899. *The Philippine Islands: A Political, Geographical, Ethnographical, Social and Commercial History of the Philippine Archipelago and Its Political Dependencies, Embracing the Whole Period of Spanish Rule*. New York: C. Scribner's Sons.

——. 1906. *The Philippine Islands*. Shanghai: Kelly and Walsh.

Forest District. 1905. "Forest District of Bataan and Zambales." In *Annual Report of the Chief of the Bureau of Forestry of the Philippine Islands for the Period September 1, 1903, to August 31, 1904*, 22–32. Manila: Bureau of Printing.

Fox, Robert B. 1979. "The Philippines in Prehistoric Times." In *Readings in Philippine Prehistory*, edited by M. Garcia, 35–61. Manila: Filipiniana Book Guild.

Frasch, Tilman. 1996. *Pagan, Stadt und Staat*. Stuttgart: Franz Steiner Verlag.

Frédéric, Louis. 1981. *La Vie Quotidienne dans la Péninsule Indochinoise à l'Époque d'Angkor, 800–1300*. Paris: Hachette.

Fukui, Hayao. ed. 1999. *The Dry Areas of Southeast Asia: Harsh or Benign Environment?* Kyoto: Center for Southeast Asian Studies.

Fuller, Douglas O. and Kevin Murphy. 2006. "The Enso-fire Dynamic in Insular Southeast Asia." *Climate Change* 74: 435–455.

Furnivall, J.S. 1939. *Netherlands India*. Cambridge [Eng.]: University Press.

Gaastra, Femme. 2003. *The Dutch East India Company: Expansion and Decline*. Zutphen: Walburg.

Gallop, Annabel Teh. 1991. *Golden Letters: Writing Traditions of Indonesia*. London: The British Library.

——. 1995. *Early Views of Indonesia/Pemandangan Indonesia di Masa Lampau: Drawings from the British Museum*. Honolulu: University of Hawai'i Press.

——. 2011. "Gold, Silver and Lapis Lazuli: Royal Letters from Aceh in the 17th Century" In *Mapping the Acehnese Past*, edited by R. Michael Feener, Patrick Daly and Anthony Reid, 105–140. Leiden: KITLV Press.

Geertz, Clifford. 1960. *The Religion of Java*. Glencoe: Free Press.

——. 1980. *Negara: The Theatre State in Nineteenth Century Bali.* Princeton: Princeton University Press.

Gentil, Guillame Le. 1903–1909. "Voyage dans les Mers d'Indie." In *The Philippine Islands, 1493–1803; Vol. 28*, edited by Emma Blair and Alexander Robertson, 192–225. Cleveland: A.H. Clark Co. Originally published 1781.

Gerini, G.E. 1905. "Historical Retrospect of Junk Ceylon Island." *Journal of the Siam Society* 2 (2): 121–268.

——. 1909. *Researches on Ptolemy's Geography of Eastern Asia*. London: Royal Asiatic Society.

Gibson, Thomas. 2000. "Islam and the Spirit Cults of New Order Indonesia: Global Flows V Local Knowledge." *Indonesia* 69 (April): 41–70.

——. 2005. *And the Sun Pursued the Moon: Symbolic Knowledge and Traditional Authority among the Makassar*. Honolulu: University of Hawai'i Press.

Giersch, C. Patterson. 2006. *Asian Borderlands: The Transformation of Qing China's Yunnan Frontier*. Cambridge, MA: Harvard University Press.

Gill, William. 1883. *The River of Golden Sand, a Journey through China and Eastern Tibet to Burmah.* London: John Murray. Second edition.

Glaser B. and M. Glaser 2010. "People, Fish and Reefs." LOICZ. Accessed 23 March 2014. http://www.loicz.org/imperia/md/content/loicz/report/People_fish_and_coral_reefs_.pdf.

Glover, Ian and Peter Bellwood. eds. 2004. *Southeast Asia: From Pre-history to History.* London and New York: Routledge Curzon.

Godley, David. 1997. "Flood Regimes in Northern Thailand." MA thesis, Monash University.

Golay, Frank et al. 1969. *Underdevelopment and Economic Nationalism in Southeast Asia*. Ithaca, NY: Cornell University Press.

Gómez-Centuríon Jimenez, Carlos. 2009. "Curiosidades Vivas: Los Animals de America y Filipinas en la Ménagerie real Durante el siglo XVIIII." *Anuario de Estudios Americanos* 66 (2): 181–211.

Gomez, C., M. Janin, F. Lavigne, R. Gertisser, S. Charbonnier, P. Lahitte, S.R. Hadmoko, M. Foert, P. Wassmer, V. Degroot and H. Murwanto. 2010. "Borobudur, a Basin under Volcanic Influence: 361,000 year BP to Present." *Journal of Volcanology and Geothermal Research* 196 (3–4): 245–264. DOI:10.1016/j.jvolgeores.2010.08.001.

Gonda, J. 1933–1936. *Agastyaparwa*. Published, translated and commented on by J. Gonda. 's-Gravenhage: n.p. Taken out from *Bijdragen tot de Taal-, Land- en Volkenkunde* 90: 329–419; 92: 337–458; 94: 223–285.

Goodall, Daphne M. 1973. *Horses of the World*. Newton Abbot: David and Charles. Third edition.

Gruber, Petra. 2007. "Adaptation and Earthquake Resistance of Traditional Nias Architecture." Nias Island Research Network. Accessed 21 July 2012. http://www.nirn.org/pdf/01_070614_gruber.pdf.

Gschwend, Annemarie Jordan. 1994. *The Development of Catherine of Austria's Collection in the Queen's Household: Its Character and Cost*. PhD thesis, Brown University.

——. 2004. "Rarities and Novelties." In *Encounters: The Meeting of Asia and Europe, 1500–1800*, edited by Anna Jackson and Amin Jaffa, 34–41. London: Victoria and Albert Museum.

Guillot, Claude. ed. 1998–2003. *Histoire de Barus*. 2 vols. Paris: Association Archipel.

Guy, John. 2011a. "Gold in the Philippines: Form, Meaning, and Metamorphosis." In *Philippine Ancestral Gold*, edited by Florina H. Capistrano-Baker, Chapter 3. Manila and Singapore: Ayala Museum and NUS Press.

——. 2011b. "Tamil Merchants and the Hindu-Buddhist Diaspora in Early Southeast Asia." In *Early Interactions between South and Southeast Asia; Reflections on Cross-cultural Exchange*, edited by Pierre-Yves Manguin, A. Mani and Geoff Wade, 243–262. Singapore: ISEAS Publishing.

Haan, F. de. 1922–1923. *Oud Batavia*. 4 vols. Vol. 3. *Platenalbum*. Batavia: Kolff.

Hadiz, Vedi and Richard Robison. 2004. *Reorganising Power in Indonesia: The Politics of Oligarchy in an Age of Markets*. London, New York: RoutledgeCurzon.

Hadler, Jeffrey. 2008. *Muslims and Matriarchs: Cultural Resilience in Minangkabau through Jihad and Colonialism*. Ithaca: Cornell University Press.

Hägerdal, Hans. 2012. *Lords of the Land, Lords of the Sea: Conflict and Adaptation in Early Colonial Timor, 1600–1800*. Leiden: KITLV Press.

Halim, Wahyuddin. 2004. "Shari'Ah Implementation in South Sulawesi: An Analysis of the Kppsi Movement." *Future Islam*. Accessed 5 July 2014. http://www.futureislam.com/20050701/insight/Wahyuddin_Halim/Shariah_Implementation_in_South_Sulawesi.asp.

Hall, D.G.E. 1955. *A History of South-East Asia*. London: Macmillan.

Hall, Kenneth R. 1992. "Economic History of Early Southeast Asia." In *The Cambridge History of Southeast Asia*, edited by Nicholas Tarling, 183–275. 4 vols. Vol. 1. Cambridge: Cambridge University Press.

———. 2004. "Local and International Trade and Traders in the Straits of Melaka Region: 600–1500." *Journal of the Economic and Social History of the Orient* 47 (2): 213–260.

———. 2011. *A History of Early Southeast Asia: Maritime Trade and Societal Development, 100–1500*. Plymouth: Rowman and Littlefield.

Hamonic, Gilbert. 1991. "God, Divinities and Ancestors: For the Positive Representation of A 'Religious Plurality' in Bugis Society, South Sulawesi, Indonesia." *Southeast Asian Studies* 29 (1): 3–34.

Hamilton, Alexander. [1727] 1930. *A New Account of the East Indies*. 2 vols. Vol. 2. Edinburgh: John Mosman; reprinted London: Argonaut Press.

Harris, Hugh. 2012. "Key to Coconut Cultivation on the American Pacific Coast: The Manila-Acapulco Galleon Route, 1565–1815." *Palms* 56 (2): 72–77.

Harvey, Barbara Sillers. 1974. *Tradition, Islam, and Rebellion: South Sulawesi 1950–1965*. PhD, Cornell University.

———. 1977. *Permesta: Half a Rebellion*. Ithaca, NY: Cornell Modern Indonesia Project.

Heidarzadeh, Mohammad, Moharram Pirooz, Nasser Zakir, Ahmet Yalciner, Mohamed Mokhtari and Asad Esmaeily. 2008. "Historical Tsunami in the Makran Subduction Zone off the Southern Coasts of Iran and Pakistan and Results of Numerical Modeling." *Ocean Engineering* 35 (8–9): 776.

Heidhues, Mary. 1992. *Bangka Tin and Mentok Pepper: Chinese Settlement on an Indonesian Island*. Singapore: ISEAS.

———. 2003. *Golddiggers, Farmers, and Traders in the "Chinese Districts" of West Kalimantan, Indonesia*. Ithaca, NY: Southeast Asia Program.

Heng, Derek. 2009. *Sino-Malay Trade and Diplomacy from the Tenth through the Fourteenth Century*. Athens, Ohio: Ohio University Press.

Henley, David. 2002. "Population, Economy and Environment in Island Southeast Asia: An Historical View with Special Reference to Northern Sulawesi." *Singapore Journal of Tropical Geography* 23 (2): 167–206.

———. 2005a. *Fertility, Food and Fever: Population, Economic and Environment in North and Central Sulawesi, 1600–1930*. Leiden: KITLV Press.

———. 2005b. "Population and the Means of Subsistence: Explaining the Historical Demography of Island Southeast Asia, with Particular Reference to Sulawesi." *Journal of Southeast Asian Studies* 36: 337–372.

———. 2006. "From Low to High Fertility on Sulawesi (Indonesia) during the Colonial Period: Explaining the 'First Fertility Transition'." *Population Studies* 60 (3): 309–327.

———. 2008. "Natural Resource Management: Historical Lessons from Indonesia." *Human Ecology* 36 (2): 273–290.

——. 2011. "Forced Labour and Rising Fertility in Colonial Indonesia." *Asian Population Studies* 7: 3–13.

Hermansyah. 2011. "The Acehnese Adaptation and Mitigation on Earthquake and Tsunami in the 18th Century (An Ancient Manuscript Research)." Paper prepared for the 3rd ICAIOS Conference, Banda Aceh, 2011.

Heshusius, C.A. 1978. *KNIL-Cavalerie, 1814–1950: Geschiedenis van de Cavalerie en Pantsertroepen van het Koninklijk Nederlands-Indische Leger*. 's-Gravenhage: Koninklijke Landmacht (KL), Sectie Krijgsgeschiedenis.

Hidalgo Nuchera, Patricio. 1995. *Encomienda, Tributo y Trabajo en Filipinas*. Madrid: Universidad Autónoma de Madrid.

Hill, Ann Maxwell. 1998. *Merchants and Migrants: Ethnicity and Trade among the Yunnanese Chinese in Southeast Asia*. New Haven: Yale University Press.

Hill, Hal. 1996. *The Indonesian Economy since 1966: Southeast Asia's Emerging Giant*. Cambridge: Cambridge University Press.

Hirth, Friedrich. 1885. *China and the Roman Orient: Researches into Their Ancient and Medieval Relations*. Shanghai and Hongkong: Kelly and Walsh.

——. 1896. "Chao Ju Kua: A New Source of Medieval Geography." *Journal of the Royal Asiatic Society of Great Britain and Ireland* 28 (1): 57–82.

Hoekstra, Pieter. 1948. *Paardenteelt op het eiland Soemba*. Batavia: Drukkerij V/H John Kappee.

Huff, Gregg. 2012. "Gateway Cities and Urbanisation in Southeast Asia before World War II." University of Oxford Discussion Papers in Economic and Social History.

Husken, Frans. 1988. *Een dorp op Java: Sociale Differentiatie in een Boerengemeenschap, 1850–1980*. Overveen: Acasea.

——. 1994. "Village Elections in Central Java; State Control or Local Democracy?" In *Leadership on Java. Gentle Hints, Authoritarian Rule*, edited by Hans Antlöv and Sven Cederroth, 119–136. Richmond: Curzon.

Hutterer, Karl L. 1974. "The Evolution of Philippine Lowland Societies." *Mankind* 9: 287–299.

Impey, Oliver and Christiaan Jörg. 2005. *Japanese Export Lacquer 1580–1850*. Amsterdam: Hotei Publishing.

Indo-China. 1943. *Indo-China*. Cambridge: Admiralty, Naval Intelligence Division.

Islam, Arshad n.d. "Arab Navigation in the Indian Ocean before European Dominance in South and Southeast Asia: A Historical Study." IIUM Repository (IREP). Accesed 17 June 2014. http://irep.iium.edu.my/3462/1/Arab_Navigation_.pdf.

Ito, Takeshi. 1984. *The World of the Adat Aceh, a Historical Study of the Sultanate of Aceh*. PhD thesis, Australian National University.

Ishii, Yoneo. 1998. *The Junk Trade from Southeast Asia*. Singapore: ISEAS, Canberra: Research School of Pacific and Asian Studies.

Iskandar, Teuku. ed. 1958. *De Hikajat Atjéh*. 's-Gravenhague: Nijhoff.

Jacq-Hergoualc'h, Michel. 1979. *L'Armement et l'Organisation de l'Armée Khmère aux XIIe et XIIIe Siècles, d'après les Bas-reliefs d'Angkor Vat, du Bàyon et de Bantlay Chmar.* Paris: Presses Universitaires de France.

Jardine, Nicolas, James A. Secord and Emma C. Spary. eds. 1996. *Cultures of Natural History.* Cambridge: Cambridge University Press.

Jocano, F. Landa. 1982. *The Ilocanos: An Ethnography of Family and Community Life in the Ilocos Region.* Quezon City: University of the Philippines.

Jomo, K.S. and Brian Folk. eds. 2003. *Ethnic Business: Chinese Capitalism in Southeast Asia.* London: RoutledgeCurzon.

Jong Boers, Bernice de. 1995. "Mount Tambora in 1815: A Volcanic Eruption in Indonesia and Its Aftermath." *Indonesia* 60: 37–59.

Jordana y Morera, Ramón. 1891. *Estudio Forestal Acerca de la India Inglesa, Java y Filipinas.* Madrid: Imprenta de Moreno y Rojas.

Jorink, Eric. 2010. *Reading the Book of Nature in the Dutch Golden Age.* Translated by Peter Mason. Leiden: Brill.

Joshi, Maheshwar P. and C.W. Brown. 1987. "Some Dynamics of Indo-Tibetan Trade through Uttarakhanda (Kumaon-Garhwal), India." *Journal of the Economic and Social History of the Orient* 30 (3): 303–317.

Junker, Laura L. 1999. *Raiding, Trading and Feasting: The Political Economy of the Philippines Chiefdoms.* Honolulu: University of Hawai'i Press.

Kaptein, Nico. 1997. *The Muhimmat al-Nafa'is: A Bilingual Meccan Fatwa Collection for Indonesian Muslims from the End of the Nineteenth Century.* Jakarta: INIS.

Karafet, Tatiana M., J.S. Lansing, Alan J. Redd, Joseph C. Watkins, S.P.K. Surata, W.A. Arthawiguna, Laura Mayer, Michael Bamshad, Lynn B. Jorde and Michael F. Hammer. 2005. "Balinese Y-chromosome Perspective on the Peopling of Indonesia: Genetic Contributions from Pre-neolithic Hunter-Gatherers, Austronesian Farmers, and Indian Traders." *Human Biology* 77-1: 93–113.

Kats, J. 1910. *Sang Hyang Kamahāyānikan: Oud-Javaansche Tekst met Inleiding, Vertaling en Aanteekeningen.* 's-Gravenhage: Nijhoff.

Kathirithamby-Wells, Jeyamalar. 1993. "Restraints on the Development of Merchant Capitalism in Southeast Asia before c. 1800." In *Southeast Asia in the Early Modern Era: Trade, Power and Belief*, edited by Anthony Reid, 123–148. Ithaca and London: Cornell University Press.

Kaufmann, Thomas DaCosta. 2010. *Arcimboldo: Visual Jokes, Natural History, and Still Life Painting.* Chicago: University of Chicago Press.

Kay, Robert. 1939. "Java Ponies and Others." *The Horse (Illustrated), The Quarterly Review of the Institute of the Horse and Pony Club* 11 (41): 24–27.

Keys, David. 2000. *Catastrophe: An Investigation into the Origins of the Modern World.* London: Random House.

Ken, Wong Lin. 1960. "The Trade of Singapore 1819–69." *Journal of the Malayan Branch, Royal Asiatic Society* 33 (4): 1–315.

Kern, H. 1885. "Sanskrit-inscriptie ter Eere van den Javaanschen Vorst Er-Langga." *Bijdragen tot de Taal- land- en Volkenkunde van Nederlandsch-Indië* 10: 1–21.

Kidd, Jane. 1985. *The Horse: The Complete Guide to Horse Breeds and Breeding*. London: Tiger Books International.

Kipling, John L. 1921. *Beast and Man in India: A Popular Sketch of Indian Animals in Their Relation with People*. London: Macmillan.

King Dwight. 1982. "Indonesia's New Order as a Bureaucratic Polity, a Neopatrimonial Regime or Bureaucratic Authoritarian Regime. What Difference Does It make?" In *Interpreting Indonesian Politics. Thirteen Contributions to the Debate*, edited by B. Anderson and A. Kahin, 104–116. Ithaca: Cornell University Press.

Klinken, Gerry van. 2009. "Patronage Democracy in Provincial Indonesia." In *Rethinking Popular Representation*, edited by O. Törnquist, N. Webster and K. Stokke, 141–159. Basingstoke: Palgrave Macmillan.

——. 2014. *The Making of Middle Indonesia. Middle Classes in Kupang Town, 1930s–1980s*. Verhandelingen KITLV 293. Leiden/Boston: Brill.

Klinken, Gerry van and Joshua Barker. 2009. "Introduction: State in Society in Indonesia." In *State of Authority: The State in Society in Indonesia*, edited by Gerry van Klinken and Joshua Barker, 1–16. Ithaca: Cornell Southeast Asia Program.

Knaap, Gerrit. 1996. *Shallow Waters, Rising Tide: Shipping and Trade in Java around 1775*. Leiden: KITLV.

——. 1999. *Cephas, Yogyakarta: Photography in the Service of the Sultan*. Leiden: KITLV Press.

Knaap, Gerrit and Heather Sutherland. 2004. *Monsoon Traders: Ships, Skippers and Commodities in Eighteenth-Century Makassar*. Leiden: KITLV.

Knaap, Gerrit and Luc Nagtegaal. 1991. "A Forgotten Trade: Salt in Southeast Asia, 1670–1813." In *Emporia, Commodities, and Entrepreneurs*, edited by Roderich Ptak and Dietmar Rothermund, 127–158. Stuttgart: Steiner.

Knapen, Han. 2001. *Forests of Fortune? The Environmental History of Southeast Borneo, 1600–1880*. Leiden: KITLV.

Kompas, 10 October 2009, www.nasional.kompas.com, 'De Haan, Heuken and 'Dari Goenoeng Sarie ke Kelenteng Sentiong.'

Kratoska, Paul H., Remco Raben and Henk Schulte Nordholt. eds. 2005. *Locating Southeast Asia: Geographies of Knowledge and the Politics of Place*. Singapore: NUS Press, Leiden: KITLV Press.

Kreemer, J. 1922–1923. *Atjèh: Algemeen Samenvattend Overzicht van Land en Volk van Atjèh en Onderhoorigheden*. Leiden: E.J. Brill.

Kulke, Hermann, K. Kesavapany and Vijay Sakhuja. eds. 2009. *Nagapattinam to Suvarnadwipa: Reflections on the Chola Naval Expeditions to Southeast Asia*. Singapore: Institute of Southeast Asian Studies.

Kumar, Ann and John H. McGlynn. eds. 1996. *Illuminations: The Writing Traditions of Indonesia*. New York: Weatherhill.

Kummer, David. 1992. *Deforestation in the Postwar Philippines*. Quezon City: Ateneo University Press.

Kusen. 1988–1989. *Factor-factor Penyebab Terjadinya Perubahan Status Sawah di Wanua Tĕngah dalam Masa Pemerintahan Raja-raja Mataram Kuna, Abad 8–10*. Yogyakarta: Fakultas Sastra Universitas Gadjah Mada.

Kusumohartono, Boegi. ed. 1993. *Temu Evaluasi Penelitian Wonoboyo, Berkala Arkeologi* xiii. Special issue.

Kwee Hui Kian. 2006. *The Political Economy of Java's Northeast Coast, c. 1740–1800*. Leiden: Brill.

——. 2007. "Pockets of Empire: Integrating the Studies on Social Organizations in Southeast China and Southeast Asia." *Contemporary Studies of South Asia, Africa, and the Middle East* 27 (3): 616–632.

——. 2008. "Cultural Strategies, Economic Dominance: The Lineage of Tan Bing in Nineteenth-Century Semarang, Java." In *Linking Destinies: Trade, Towns and Kin in Asian History*, edited by Peter Boomgaard, Dick Kooiman and Henk Schulte Nordholt, 197–217. Verhandelingen KITLV 256. Leiden: KITLV Press.

——. 2011. "The End of the 'Age of Commerce'? Javanese Cotton Trade Industry from the Seventeenth to the Eighteenth Centuries." In *Chinese Circulations: Capital, Commodities, and Networks in Southeast Asia*, edited by Eric Tagliacozzo and Wen-Chin Chang, 283–302. Durham, NC: Duke University Press.

——. 2013. "Chinese Economic Dominance in Southeast Asia: A Longue Duree Perspective." *Comparative Studies in Society and History* 55 (1): 5–34.

Laarhoven, Ruurdje. 1994. *The Power of Cloth: The Textile Trade of the Dutch East India Company (VOC) 1600–1780*. PhD thesis, Australian National University.

Lach, Donald F. and Edwin J. van Kley. 1993. *Asia in the Making of Europe*. 3 vols. Vol.3. Chicago: University of Chicago Press.

Lande, Carl. 1964. *Leaders, Factions, and Parties: The Structure of Philippine Politics*. New Haven: Yale University SEA Studies.

Lansing, J.S., A.J. Redd, T.M. Karafet, J. Watkins, I.W. Ardika, S.P.K. Surata, J.S. Schoenfelder, M. Campbell, A.M. Merriwether and M.F. Hammer. 2004. "An Indian Trader in Ancient Bali?" *Antiquity* 78 (300): 287–293.

Lasco, Rodel D. and Juan M. Pulhin. 2006. "Environmental Impacts of Community-Based Forest Management in the Philippines." *International Journal of Environment and Sustainable Development* 5: 46–56.

Lavigne, Franck, Jean-Philippe Degeai et al. 2013. "Source of the Great A.D. 1257 Mystery Eruption Unveiled, Samalas Volcano, Rinjani Volcanic Complex, Indonesia." *Proceedings of the National Academy of Sciences of the United States of America* 110 (42): 16742–16747.

Ledesma, Charles de, Lewis, Mark and Savage, Pauline. 1997. *Malaysia, Singapore and Brunei: The Rough Guide*. London: Rough Guides.

Lee Kam Hing. 1995. *The Sultanate of Aceh: Relations with the British, 1760–1824*. Kuala Lumpur: Oxford University Press.

Legendre, A.-F. 1913. *Au Yunnan et dans le Masif du Kin-Ho, Fleuve d'Or*. Paris: Plon.

Leopold, Aldo. 1979. "Some Fundamentals of Conservation in the Southwest." *Environmental Ethics* 1: 131–148.

Leur, J.C. van. 1955. *Indonesian Trade and Society: Essays in Asian Social and Economic History*. The Hague: W. Van Hoeve.

——. 1960. *Indonesian Trade and Society: Essays in Asian Social and Economic History*. Bandung: Sumur Bandung. Second edition.

Lévi-Strauss, Claude. 1958–1973. *Anthropologie Structurale*. 2 vols. Paris: Plon.

——. [1949] 1968. "Introduction: History and Anthropology." *Structural Anthropology* I: 1–27. Harmondsworth: Penguin.

Lewis, Dianne. 1995. *Jan Compagnie in the Straits of Malacca 1641–1795*. Athens: Ohio University Center for International Studies.

Li, Tana. 1998. *Nguyen Cochinchina: Southern Vietnam in the Seventeenth and Eighteenth Centuries*. Ithaca, NY: Southeast Asia Program Publications.

——. 2011. "Cochinchinese Coin Casting and Circulating in Eighteenth-Century Southeast Asia." In *Chinese Circulations: Capital, Commodities, and Networks in Southeast Asia*, edited by Eric Tagliacozzo and Wen-Chin Chang, 130–148. Durham, NC: Duke University Press.

Li Tana and Anthony Reid. eds. 1993. *Southern Vietnam under the Nguyen: Documents on the Economic History of Cochinchina (Dang Trong), 1602–1777*. Singapore: Institute of Southeast Asian Studies.

Lieberman, Victor. 1993. "Local Integration and Eurasian Analogies: Structuring Southeast Asian History." *Modern Asian Studies* 27 (3): 478–572.

——. 1995. "An Age of Commerce in Southeast Asia? Problems of Regional Coherence – A Review Article." *Journal of Asian Studies* 54 (3): 796–807.

——. 1999. *Beyond Binary Histories: Re-imagining Eurasia to c.1830*. Ann Arbor: University of Michigan Press.

——. 2003. *Strange Parallels: Southeast Asia in Global Context, c. 800–1830*. 2 vols. Vol. 1. *Integration of the Mainland*. Cambridge: Cambridge University Press.

——. 2009. *Strange Parallels: Southeast Asia in Global Context, c. 800–1830*. 2 vols. Vol. 2. *Mainland Mirrors: Europe, Japan, China, South Asia and the Islands*. Cambridge: Cambridge University Press.

——. 2010. "Maritime Influences in Southeast Asia, c. 900–1300: Some Further Thoughts." *Journal of Southeast Asian Studies* 41 (3): 529–539.

Loeb, Edwin. 1935. *Sumatra: Its History and People*. Vienna: Institut für Völkerkunde der Universität Wien.

Lindblad, J. Thomas. 1998. *Foreign Investment in Southeast Asia during the Twentieth Century*. Houndmills, Basingstoke, Hampshire: Macmillan.

Lombard, Denys. 1967. *Le Sultanat d'Atjéh au Temps d'Iskandar Muda, 1607–1636*. Paris: École Française d'Extrême Orient.

Loos-Haaxman, J. de. 1941. *De Landsverzameling Schilderijen in Batavia: Landsvoogds-portretten en Compagnieschilders*. Leiden: Sijthoff.

Loureiro, Rui M. 2009. "Portuguese Involvement in the Sixteenth-Century Horse-Trade through the Arabian Sea." In *Pferde in Asien: Geschichte, Handel und Kultur*, edited by Bert Fragner, Ralph Kauz, Roderich Ptak and Angela Schottenhammer, 137–143. Vienna: Verlag der Österreichischen Akademie der Wissenschaften.

Lubinsky, Pesach, Kenneth M. Cameron, María Carmen Molina, Maurice Wong, Sandra Lepers-Andrzejewski, Arturo Gómez-Pompa and Seung-Chul Kim. 2008. "Neotropical Roots of a Polynesian Spice: The Hybrid Origin of Tahitian Vanilla, *Vanilla Tahitensis* ORCHIDACEAE." *American Journal of Botany* 95 (8): 1040–1047.

Luce, G.H. 1960. "Economic Life of the Early Burmans." *Burma Research Society fiftieth Anniversary Publications* 2: 323–375. Rangoon: Burma Research Society.

Lund, Christian. 2011. "Property and Citizenship: Conceptually Connecting Land Rights and Belonging in Africa." *Africa Spectrum* 46: 71–75.

Lustig, Eileen. 2009. *Power and Pragmatism in the Political Economy of Angkor*. PhD thesis, University of Sydney.

Ma Huan. 1970. *Ying-yai Sheng-lan: The Overall Survey of the Ocean's Shores*. Cambridge: Cambridge University Press.

MacKnight, C.C. 1983. "The Rise of Agriculture in South Sulawesi before 1600." *Review of Indonesian and Malaysian Affairs* 17: 92–116.

Magenda, Burhan D. 1989. *The Surviving Aristocracy in Indonesia: Politics in Three Provinces of the Outer Islands*. PhD thesis, Cornell University.

Majumdar, R.C. 1963. *Hindu Colonies in the Far East*. Calcutta: K.L. Mukhopadhyay. Second edition.

Mangemba, H.D. 1972. *Kota Makassar: Dalam Lintasan Sedjarah*. Makassar: Lembaga Sejarah Fakultas Sastra Universitas Hasanuddin.

Manguin, Pierre-Yves. 2011. "Introduction." In *Early Interactions between South and Southeast Asia; Reflections on Cross-Cultural Exchange*, edited by Pierre-Yves Manguin, A. Mani and Geoff Wade, xiii–xxxi. Singapore: Institute of Southeast Asian Studies.

Manguin, Pierre-Yves, A. Mani and Geoff Wade. eds. 2011. *Early Interactions between South and Southeast Asia: Reflections on Cross-Cultural Exchange*. Singapore: Institute of Southeast Asian Studies.

Manning, Patrick. 2003. *Navigating World History: Historians Create a Global Past*. New York: Palgrave Macmillan.

Marcy, Peter T. 1981. "Factors Affecting the Fecundity and Fertility of Historical Populations." *Journal of Family History* 6: 309–326.

Marsden, William. 1966. *A History of Sumatra*. Kuala Lumpur: OUP. Third revised edition.

Martín-Meras, María Luisa and María Dolores Higueras. eds. 1974. *La Historia de las Islas e Indios del Padre Alcina 1668*. Madrid: Instituto Histórico de Marina.

Mathers, William M., Henry S. Parker III and Kathleen A. Copus. eds. 1990. *Archaeological Report: The Recovery of the Manila Galleon Nuestra Señora de la Concepción*. Sutton, Vermont: Pacific Sea Resources.

Matthews, Donald. 1914. *Silviculture for the Philippines*. Los Baños: College of Agriculture, University of the Philippines.

Mattulada. 1976. *Agama Islam Di Sulawesi Selatan: Laporan Proyek Penelitian Peranan Ulama Dan Pengajaran Agama Islam Di Sulawesi Selatan*. Makassar: Universitas Hasanuddin Fakultas Sastra.

Mazlish, Bruce. 1998. "Comparing Global History to World History." *Journal of Interdisciplinary History* 28 (3): 385–395.

———. 2001. "On history Becoming History: The Case of World and New Global History." http://www.newglobalhistory.com/docs/mazlish-on-history-becoming-history.pdf.

McCaffrey, Robert. 2009. "The Tectonic Framework of the Sumatran Subduction Zone." *Annual Review of Earth and Planetary Sciences* 37: 345–366. http://www.annualreviews.org/doi/abs/10.1146/annurev.earth.031208.100212.

McGee, Terry G. 1991. "The Emergence of Desakota Regions in Asia: Expanding a Hypothesis." In *The Extended Metropolis: Settlement Transition in Asia*, edited by Norton Sydney Ginsburg, Bruce M. Koppel and T.G McGee, 3–25. Honolulu: University of Hawaii Press.

Meilink-Roelofsz, M.A.P. 1962. *Asian Trade and European Influence in the Indonesian Archipelago between 1500 and about 1630*. The Hague: Martinus Nijhoff.

Metzner, Aron, Kerry Sieh, Hong-Wei Chiang, Chuan-Chou Shen, Belle E. Philibosian, Bambang W. Suwargadi and Danny H. Natawidjaja. 2009. "Coral Microatoll Paleogeodesy on Simeulue Island, Sumatra, reveals Earthquake Clusters and Persistent Rupture Segmentation." Paper presented at Portland meeting of the Geological Society of America, GSA *Abstracts with Programs* 41 (7): 409.

Meltzner, A.J., K. Sieh, H.-W. Chiang, C.-C. Shen, B.W. Suwargadi, D.H. Natawidjaja, B.E. Philibosian, R.W. Briggs and J. Galetzka. 2010. "Coral Evidence for Earthquake Recurrence and an A.D. 1390–1455 Cluster at the South end of the 2004 Aceh-Andaman Rupture." *Journal of Geophysical Research* 115, B10. DOI:10.1029/2010JB007499.

Milner, Anthony. 1982. *Kerajaan: Malay Political Culture on the Eve of Colonial Rule*. Tucson: University of Arizona Press.

Miller, Hugo. 1911. *Commercial Geography: The Materials of Commerce for the Philippines*. Manila: Bureau of Printing.

Mirandola, Andrés de. 1903–1909. "Letter to Felipe II." In *The Philippine Islands, 1493–1803; Vol. 3*, edited by Emma Blair and Alexander Robertson, 223–229. Cleveland: A.H. Clark Co. Originally published 1574.

Miksic, John. 2009. "Highland-Lowland Connections in Jambi, South Sumatra, and West Sumatra, 11th to 14th Centuries." In *From Distant Tales: Archeology and Ethnohistory in the Highlands of Sumatra,* edited by Dominik Bonatz, John Miksic, J. David Neidel and Mai Lin Tjoa-Bonatz, 75–102. Cambridge: Cambridge Scholars Publishing.

Monecke, Katrin, Willi Finger, David Klarer, Widjo Kongko, Brian McAdoo, Andrew Moore and Sam Sudrajat. 2008. "A 1,000-Year Sediment Record of Tsunami Recurrence in Northern Sumatra." *Nature* 455: 1232–1234.

Montemayor, M.R. 1952. "Half a Century of Livestock Raising." In *A Half Century of Philippine Agriculture*. Manila: Graphic House.

Moor, J.H. comp. 1968. *Notices of the Indian Archipelago and Adjacent Countries.* London: Frank Cass. Reprint of 1837.

Morga, Antonio de. 1903–1909. "Sucesos de las Islas Filipinas." In *The Philippine Islands, 1493–1803; Vol. 16,* edited by Emma Blair and Alexander Robertson, 25–210. Cleveland: A.H. Clark Co. Originally published 1609.

———. 1971. *Sucesos de las Islas Filipinas,* edited by James S. Cummins. Vol. 140. Cambridge: Cambridge University Press. Hakluyt Society, 2nd Series.

Mortimer, Rex. 1974. *Indonesian Communism under Sukarno: Ideology and Politics, 1959–1965.* Ithaca: Cornell University Press.

Muijzenberg, Otto van de. 2000. *Four Centuries of Dutch-Philippine Economic Relations 1600–2000.* Manila: Royal Netherlands Embassy.

Mundardjito, O. 1978. *Preliminary Report of the Pottery found at the Borobudur Site.* Jakarta: Badan Pemugaran Candi Borobodur.

Mundy, Peter. 1919. *The Travels of Peter Mundy.* London: Hakluyt Society.

Mus, Paul. 1933. "L'Inde vue de l'Est: Cultes Indiens et Indigènes au Champa." *Bulletin de l'Ecole française d'Extrême-Orient* 33 (1): 367–410.

———. 2011. *India Seen from the East: Indian and Indigenous Cults in Champa,* edited by Ian Mabbett and David Chandler. Caulfeld: Monash University Press.

Naess, Arne. 1973. "The Shallow and the Deep, Long-Range Ecology Movement: A Summary." *Inquiry* 16: 95–100.

Nagtegaal, Luc. 1996. *Riding the Dutch Tiger: The Dutch East Indies Company and the Northeast Coast of Java, 1680–1743.* Leiden: KITLV Press.

Nakada, K. 1990. "A Palaeographical Study of Indonesian Inscriptions (III)." *Historical Science Reports, Kagoshima University* 37: 2–21.

Nano, Jose. 1953. "Brief History of Forestry in the Philippines." *Philippine Journal of Forestry* 8 (1–4): 9–128.

Nash, Manning. 1965. *The Golden Road to Modernity. Village Life in Contemporary Burma.* New York: Wiley.

Nash, Roderick. 1972. "American Environmental History: A New Teaching Frontier." *Pacific Historical Review* 41: 362–372.

Nastiti, Titi Surti. 2004. *Prasasti Sobhāmṛta.* Unpublished manuscript.

Nastati, Titi Surti, Dyah Wijaya Dewi and Richardiana Kartakusuma. 1982. *Tiga Prasasti dari Masa Balitung.* Jakarta: Pusat Penelitian Arkeologi Nasional.

Newcomb, K.R. and W.R. McCann. 1987. "Seismic History and Seismotectonics of the Sunda Arc." *Journal of Geophysical Research* 92 (B1): 421–439.

Newhall, C.G., S. Broto, B. Alloway, N.G. Banks, I. Bahar, M.A. del Marmol, R.D. Hadisantono, R.T. Holcomb, J. McGeehin, J.N. Miksic, M. Rubin, S.D. Sayudi, R. Sukhyar, S. Andreastiti, R.I. Tilling, R. Torley, D. Trimble and A.D. Wirakusumah. 2000. "10,000 Years of Explosive Eruptions of Merapi Volcano, Central Java: Archaeological and Modern Implications." *Journal of Volcanology and Geothermal Research* 100: 9–50.

Newson, Linda A. 1995. *Life and Death in Early Colonial Ecuador.* Norman: University of Oklahoma Press.

——. 1998. "A Historical-Ecological Perspective on Epidemic Disease." In *Advances in Historical Ecology*, edited by William Balée, 41–63. New York: Columbia University Press.

——. 2006. "The Demographic Impact of Colonization." In *The Cambridge Economic History of Latin America, Vol. 1: The Colonial Era and the Short Nineteenth Century*, edited by Victor Bulmer-Thomas, John H. Coatsworth, Roberto Cortés-Conde, 143–184. Cambridge: Cambridge University Press.

——. 2009. *Conquest and Pestilence in the Early Spanish Philippines.* Honolulu: University of Hawai'i Press.

Nguyen, Thanh Nha. 1965. *Tableau Économique du Vietnam aux XVIIe et XVIIIe Siècles.* Paris: Bibliothèque Bich-Thanh-Thu.

Nishimura, Masao. 1999. "The Spanish Colonisation and the Change of Landscape of Cebu Island." *Waseda Journal of Asian Studies* 21: 29–55.

Noordegraaf, Leo. 2006. "In Gesprek met Peter Boomgaard." In *Waarover Spraken Zij? Economische Geschiedbeoefening in Nederland omstreeks het Jaar 2000*, edited by Leo Noordegraaf, 45–56. Amsterdam: Aksant.

Noorduyn, J. 1983. "De Handelsrelaties van het Makassaarse Rijk volgens de Notitie van Cornelis Speelman (1669)." In *Nederlands Historische Bronnen*, 97–123. 's-Gravenhage: Martinus Nijhoff.

Nordin Hussin. 2007. *Trade and Society in the Straits of Melaka.* Singapore: NUS Press; Copenhagen: NIAS Press.

North, Michael. 1997. *Art and Commerce in the Dutch Golden Age.* New Haven: Yale University Press.

——. 2010. "Production and Reception of Art through European Company Channels in Asia." In *Artistic and Cultural Exchanges between Europe and Asia, 1400–1900*, edited by Michael North, 89–107. Aldershot: Ashgate.

North, Michael and David Ormrod. eds. 1998. *Art Markets in Europe, 1400–1800.* Aldershot: Ashgate.

Nuru'd-din ar-Raniri. 1966. *Bustanu's-Salatin, Bab II, Fasal 13*, edited by T. Iskandar. Kuala Lumpur: Dewan Bahasa dan Pustaka.

Ogilvie, Brian W. 2006. *The Science of Describing: Natural History in Renaissance Europe*. Chicago: University of Chicago Press.

Olivier, J. 1834–1837. *Reizen in den Molukschen Archipel naar Makassar, ...In het Gevolg van den Gouverneur-Generaal van Nederland's Indië in 1824 gedaan en volgens de Dagboeken en Aanteekeningen van Ondersch. Reisgenooten beschreven*. 2 vols. Amsterdam: Beijerinck.

Ormeling, J. 1956. *The Timor Problem: A Geographical Interpretation of an Underdeveloped Island*. Groningen: J.B. Wolters.

Ota, Atsushi. 2006. *Changes of Regime and Social Dynamics in West Java: Society, State, and the Outer World of Banten, 1750–1830*. Leiden: Brill.

Owen, Norman G. 1987a. "The Paradox of Nineteenth-Century Population Growth in Southeast Asia: Evidence from Java and the Philippines." *Journal of Southeast Asian Studies* 18 (1): 45–97.

——. 1987b. "Towards a History of Health in Southeast Asia." In *Death and Disease in Southeast Asia: Explorations in Social, Medical and Demographic History*, edited by Norman G. Owen, 3–30. Oxford: Oxford University Press.

——. 2002. "Population and Society in Southeast Asia before 1900." Paper presented at the IUSSP Regional Population Conference, Bangkok.

Pameran Seabad. 1992. *Pameran Seabad Kearsipan: Menunjang Pembangunan Bangsa Indonesia 1892–1992*. Jakarta. Arsip Nasional Republik Indonesia.

Parimartha, I Gde. 1995. *Perdagangan dan Politik di Nusa Tenggara, 1815–1915*. PhD thesis, Vrije Universiteit, Amsterdam.

Patji, Abdul Rachman. 1991. "The Arabs of Surabaya: A Study of Sociocultural Integration." MA Thesis, Australian National University.

Payen, Antoine A.J. 1988. *Journal de mon Voyage à Jocja Karta en 1825*. Paris: Association Archipel.

Pedrosa, Ramón. 1983. "Abortion and Infanticide in the Philippines during the Spanish Contact." *Philippiniana Sacra* 18 (52): 7–37.

Peissel, Michel. 2000. *The Last Barbarians, the Discovery of the Sources of the Mekong in Tibet*. London: Souvenir Press.

Pelras, Christian. 1985. "Religion, Tradition and the Dynamics of Islamisation in South Sulawesi." *Archipel* 29: 107–136.

——. 1994. "Religion, Tradition and the Dynamics of Islamization in South Sulawesi." *Indonesia* 57: 133–154.

——. 1996. *The Bugis*. Oxford: Blackwell.

——. 2000. "Patron-Client Ties among the Bugis and Makassarese of South Sulawesi." *Bijdragen tot de Taal-, Land- en Volkenkunde* 156: 393–432.

Pemberton, John. 1986. "Notes on the 1982 General Election in Solo." *Indonesia* 41: 1–22.

Perret, Daniel and Heddy Surachman. 2011. "South Asia and the Tapanuli Area (North-West Sumatra): Ninth-Fourteenth Centuries CE." In *Early Interactions between South and Southeast Asia; Reflections on Cross-Cultural Exchange*, edited by A. Mani Pierre-Yves Manguin and Geoff Wade, 161–175. Singapore: Institute of Southeast Asian Studies.

Perkins, Sid. 2008. "Reef Record Suggests Impending Sumatra Quakes." *ScienceNews*. Accessed 7 July 2014. https://www.sciencenews.org/article/reef-record-suggests-impending-sumatra-quakes.

Phalgunadi, I. Gusti Putu. 1996. *The Pararaton: A Study of the Southeast Asian Chronicle*. New Delhi: Sundeep Prakashan.

Phelan, John L. 1959. *The Hispanization of the Philippines: Spanish Aims and Filipino Responses 1565–1700*. Madison: University of Wisconsin.

Pigafetta, Antonio. 1995. *The First Voyage around the World: An Account of Magellan's Expedition*, edited by Theodore J. Cachey Jr. New York: Marsillo Publishers.

Pineda, Sebastian de. 1903–1909. "Philippine Ships and Shipbuilding." In *The Philippine Islands, 1493–1803; Vol. 18*, edited by Emma Blair and Alexander Robertson, 169–188. Cleveland: A.H. Clark Co. Originally published 1619.

Pinker, Steven. 2011. *The Better Angels of Our Nature: The Decline of Violence in History and Its Causes*. London: Allen lane.

Pirazzoli-t'Serstevens, Michèle. 1974. *La Civilisation du Royaume Dian à l'Époque Han*. Paris: École française d'Extrême-Orient.

Pires, Tomé. 1944. *The Suma Oriental of Tomé Pires, and the Book of Francisco Rodrigues*. London: Hakluyt Society.

Plasencia, Juan de. 1892. "Relación del culto que los Tagalogs Tenían [1676]." In *Cronica de la Provincia de San Gregorio Magno de Religiosos Descalzos de N.S.P. San Francisco en las Islas Filipinas, China, Japon, etc.*, edited by Francisco de Santa Inés, 592–603. *Biblioteca Histórica Filipina*. vol. 2, part 3. Manila: Tipo-litografia de Chofre comp.

Platt, Peter G. ed. 1999. *Wonders, Marvels, and Monsters in Early Modern Culture*. Newark: University of Delaware Press.

Ploeg, Jan van der, Merlijn Weerd, Andres B. Maspiqueña and Gerard Persoon. 2011. "Illegal Logging in the Northern Sierra Madre Natural Park, the Philippines." *Conservation and Society* 9: 202–215.

Poelinggomang, Edward L. 2002. *Makassar Abad Xix: Studi Tentang Kebijakan Perdagangan Maritim*. Jakarta: Gramedia.

———. 2004. *Perubahan Politik Dan Hubungan Kekuasaan: Makassar, 1906–1942*. Tegalrejo, Yogyakarta: Ombak.

Poerbatjaraka, R.M.Ng. 1941. "Strophe 14 van de Sanskrit-zijde der Calcutta-oorkonde." *Tijdschrift voor Indische Taal-, Land- en Volkenkunde* 81: 424–437.

Pollock, Sheldon. 2006. *The Language of the Gods in the World of Men; Sanskrit, Culture, and Power in Premodern India*. Berkeley: University of California Press.

Pomata, Gianna and Nancy G. Siraisi. 2005. *Historia: Empiricism and Erudition in Early Modern Europe.* Cambridge: MIT Press.

Pombejra, Dhiravat na. 1993. "Ayutthaya at the End of the Seventeenth Century: Was There a Shift to Isolation?" In *Southeast Asia in the Early Modern Era: Trade, Power and Belief*, edited by Anthony Reid, 250–272. Ithaca, NY: Cornell University Press.

——. 2007. "Javanese Horses for the Court of Ayutthaya." In *Breeds of Empire: The 'Invention' of the Horse in Southeast Asia and Southern Africa, 1500–1950*, edited by Greg Bankoff and Sandra Swart, 65–81. Copenhagen: NIAS Press.

Postma, A. 1991. "The Laguna Copper-Plate Inscription: A Valuable Philippine Document." In *Indo-Pacific Prehistory 1990*, edited by P. Bellwood, 160–172. 2 vols. Vol. 2. Canberra: Australian National University.

Pourias, Emile-René. 1889. *La Chine: Huit Ans au Yun-nan, récit d'un Missionaire.* Lille: Société de Saint-Augustin.

Ptak, Roderich. 1991. "Pferde auf See, ein Vergessener Aspekt des Maritimen Chinesischen Handels im frühen 15 Jahrhundert." *Journal of the Economic and Social History of the Orient* 34 (2): 199–233.

——. 1992. "The Northern Trade Route to the Spice Islands: South China Sea – Sulu zone – North Moluccas 14th to Early 16th Century." *Archipel* 43: 27–56.

——. 1993. "China and the Trade in Cloves, c. 960–1435." *Journal of the American Oriental Society* 113 (1): 1–13.

——. 1999. "Asian Trade in Cloves, circa 1500: Quantities and Trade Routes – A Synopsis of Portuguese and Other Sources." In *China's Seaborne Trade with South and Southeast Asia 1200–1750*, edited by Roderich Ptak, 149–169. Ashgate: Ashgate Variorum.

Putnam, Robert D., Robert Leonardi and Raffaella Y. Nanetti. 1992. *Making Democracy Work: Civic Traditions in Modern Italy.* Princeton: Princeton University Press.

Quaritch Wales, Horace G. 1952. *Ancient South-East Asian Warfare.* London: Bernard Quaritch.

Quiason, Serafin D. 1966. *English 'Country Trade' with the Philippines, 1644–1765.* Quezon City: University of the Philippines Press.

Quirino, Carlos and Mauro Garcia. 1958. "The Manners, Customs and Beliefs of the Philippine Inhabitants of Long Ago." *The Philippine Journal of Science* 87 (4): 325–453.

Raffles, Sir Thomas Stamford. 1830. *The History of Java.* 2 vols. Vol. 2. London: J. Murray.

Ramos, Miguel and Maria da Conceição Rodrigues. 1980. "Espólios Sepulcrais Timorenses." *Leba, Estudos de Pré-Historia e Arqueologia* 3: 47–57.

Recopilación. 1943. *Recopilación de los Leyes de los Reynos de las Indias.* 3 vols. 1943. Madrid: Gráficas Ultra.

Read, Bernard E. 1931. *Chinese Materia Medica.* Peking: Peking Natural History Bulletin. No pagination.

Reid, Anthony. 1980. "The Structure of Cities in Southeast Asia 15th to 17th Centuries." *JSEAS* XI (2): 235–258.

——. 1983. "Introduction: Slavery and Bondage in Southeast Asian History." In *Slavery, Bondage and Dependency in Southeast Asia*, edited by Anthony Reid, 1–43. New York: St. Martin's Press.

——. 1987. "Low Population Growth and Its Causes in Pre-colonial Southeast Asia." In *Death and Disease in Southeast Asia: Explorations in Social, Medical and Demographic History*, edited by Norman Owen, 33–47. Oxford: Oxford University Press.

——. 1988. *Southeast Asia in the Age of Commerce 1450–1680*. Vol. 1. *The lands below the winds*. New Haven/London; Yale University Press.

——. 1988–1993. *Southeast Asia in the Age of Commerce 1450–1680*. 2 vols. New Haven: Yale University Press.

——. 1990a. "An 'Age of Commerce' in Southeast Asian History." *Modern Asian Studies* 24 (1): 1–30.

——. 1990b. "The Seventeenth-Century Crisis in Southeast Asia." *Modern Asian Studies* 24 (4): 639–659.

——. 1993. *Southeast Asia in the Age of Commerce 1450–1680*. 2 vols. Vol. 2*: Expansion and Crisis*. New Haven and London: Yale University Press.

——. 1994. "The Unthreatening Alternative: Chinese Shipping in Southeast Asia 1567–1842." *Review of Indonesian and Malaysian Affairs* 27: 13–32.

——. 1996. "Flows and Seepages in the Long-Term Chinese Interaction with Southeast Asia." In *Sojourners and Settlers: Histories of Southeast Asia and the Chinese*, edited by Anthony Reid, 15–50. St. Leonards, NSW: Allen & Unwin.

——. 1997. "A New Phase of Commercial Expansion in Southeast Asia, 1760–1850." In *The Last Stand of Asian Autonomies*, edited by Anthony Reid, 57–81. London: Macmillan Press, New York: St Martin's Press.

——. 1998a. "Inside-Out: The Colonial Displacement of Sumatra's Population." In *Paper Landscapes: Essays in the Environmental History of Indonesia*, edited by Peter Boomgaard, Freek Colombijn and David Henley, 61–89. Leiden: KITLV Press.

——. 1998b. "Introduction: Four Key Exports and the Trade Cycle in Southeast Asia." In *Southeast Asian Exports since the 14th Century: Cloves, Pepper, Coffee, and Sugar*, edited by David Bulbeck, Anthony Reid, Lay Cheng Tan and Yiqi Wu, 1–16. Singapore: Institute of Southeast Asian Studies.

——. 2000a. "A Great Seventeenth Century Indonesian Family: Matoaya and Pattingalloang of Makassar." In *Charting the Shape of Early Modern Southeast Asia*, edited by Anthony Reid, 126–154. Singapore: Institute of Southeast Asian Studies.

——. 2000b. "Pluralism and Progress in Seventeenth-Century Makassar." In *Authority and Enterprise among the Peoples of South Sulawesi*, edited by Roger Tol, Kees van Dijk and Greg Acciaioli, 55–71. Leiden: KITLV Press.

——. 2000c. *Charting the Shape of Early Modern Southeast Asia.* Singapore: Silkworm Books/Institute of Southeast Asian Studies.

——. 2001. "South-East Asian Population History and the Colonial Impact." In *Asian Population History*, edited by Liu Ts'ui-jung, James Lee, David Sven Reher, Osamu Saito and Wang Feng, 45–62. Oxford: Oxford University Press.

——. 2006. "Introduction." In *Verandah of Violence: The Background to the Aceh Conflict*, edited by Anthony Reid, 7–8. Singapore: Singapore University Press.

——. 2010. *To Nation by Revolution: Indonesia in the 20th Century*. Singapore: NUS Press.

——. 2011. "Chinese on Mining Frontier in Southeast Asia." In *Chinese Circulations: Capital, Commodities and Networks in Southeast Asia*, edited by Eric Tagliacozzo and Chang Wen-chin, 21–36. Durham: Duke University Press.

——. 2012. "Historical Evidence for Past Tsunamis in the Java Subduction Zone." *Asia Research Institute Working Paper* 178. Asia Research Institute, National University of Singapore. www.ari.nus.edu.sg/docs/wps/wps12_178.pdf.

Reid, Anthony and Raden Fernando. 1996. "Shipping on Melaka and Singapore as an Index of Growth, 1760–1840." *South Asia* 19: 59–84.

Rémond, P. and P. Winter. 1944. *Contribution à l'Alimentation du Bétail en Indochine*. Hanoi: Gouvernement Général de l'Indochine.

Report of the Chief. 1906. *Report of the Chief of the Bureau of Forestry, 1905*. Washington: Government Printing Office.

Reyes, Raquel A.G. 1999. "Botany and Zoology in the Late Seventeenth Century Philippines: The Work of Georg Josef Camel, SJ 1661–1706." *Archives of Natural History* 36 (2): 262–276.

——. 2013. "Collecting and the Pursuit of Scientific Accuracy: The Malaspina Expedition in the Philippines, 1792." In *Empire and Science in the Making: Dutch Colonial Scholarship in Comparative Global Perspective*, edited by Peter Boomgaard, 63–89. New York: Palgrave-Macmillan.

Ricklefs, M.C. 1974. *Jogjakarta under Sultan Mangkubumi, 1749–1792*. London: Oxford University Press.

——. 1978. *Modern Javanese Historical Tradition: A Study of an Original Kartasura Chronicle and Related Materials*. London: School of Oriental and African Studies.

——. 2006. *Mystic Synthesis in Java: A History of Islamization from the Fourteenth to the Early Nineteenth Centuries*. White Plains, New York: EastBridge.

——. 2007. *Polarising Javanese Society: Islamic and Other Visions, 1830–1930*. Singapore: NUS Press.

Rijksbureau voor Kunsthistorische Documentatie, www.rkd.nl.

Robequain, Charles. 1944. *The Economic Development of French Indo-China*. London: Oxford University Press.

Robison, Richard. 1986. *Indonesia, The Rise of Capital*. Sydney: Allen and Unwin.

Robson, Stuart O. 1995. *Desawarnana (Nagarakrtagama*. Leiden: KITLV Press.

Rockhill, W.W. 1915. "Notes on the Relations and Trade of China with the Eastern Archipelago and the Coasts of the Indian Ocean during the Fourteenth Century." *T'oung Pao* ser. 2 (16): 148–149.

Rodríguez, Isacio R. 1965–1988. *Historia de la Provincia Agustiniana del Smo. Nombre de Jésus de Filipinas*. 20 vols. Manila: Arnoldus.

Roth, Denis. 1983. "Philippine Forests and Forestry: 1565–1920." In *Global Deforestation and the Nineteenth-Century World Economy*, edited by Richard Tucker and John Richards, 30–49. Durham: Duke Press.

Salmon, Claudine and Roderich Ptak. eds. 2005. *Zheng He: Images and Perceptions*. Wiesbaden: Harrassowitz Verlag.

San Agustín, Gaspar de. 1975. *Conquistas de las Islas Filipinas*. Madrid: Consejo de Investigaciones Superiores.

Sande, Francisco de. 1903–1909. "Relation of the Filipinas Islands." In *The Philippine Islands, 1493–1803; Vol. 4*, edited by Emma Blair and Alexander Robertson, 21–135. Cleveland: A.H. Clark Co. Originally published 1576.

Santosa, H.B. Herry. 1994. "Prasasti Tulang Er, 198 Sanjaya." *Berkala Arkeologi* 14: 17–29. Special edition.

Sarkar, H.B. 1971. *Corpus of the Inscriptions of Java (up to 928 A.D.)*. Vol. 1. Calcutta: Firma K.L. Mukhopadhyay.

——. 1972. *Corpus of the Inscriptions of Java (up to 928 A.D.)*. Vol. 2. Calcutta: Firma K.L. Mukhopadhyay.

Scalliet, Marie-Odette, Koos van Brakel, David van Duuren and Jeannette ten Kate. 1999. *Pictures from the Tropics*. Amsterdam: KIT.

Schafer, Edward H. 1967. *The Vermilion Bird: T'ang Images of the South*. Berkeley: University of California Press.

——. 1985. *The Golden Peaches of Samarkand: A Study of T'ang Exotics*. Berkeley: University of California Press.

Schlegel, G. 1901. "Geographical Notes XVI: The Old States in the Island of Sumatra." *T'oung Pao* ser. 2 (2): 353–359.

Schneider, Edwin. 1916. *Commercial Woods of the Philippines: Their Preparation and Uses*. Manila: Bureau of Printing.

Schoeman, Karel. 2006. *Kinders van die Kompanjie: Kaapse Lewens uit die Sewentiende Eeu*. Pretoria: Protea Boekhuis.

Schrieke, B. 1955–1957. *Indonesian Sociological Studies*. 2 vols. The Hague: W. van Hoeve.

Schulte Nordholt, Henk. 1993. "Leadership and the Limits of Political Control: A Balinese 'Response' to Clifford Geertz." *Social Anthropology* 1: 291–307.

——. 1996. *The Spell of Power: A History of Balinese politics, 1650–1940*. Leiden: KITLV Press.

——. 2002. "A Genealogy of Violence." In *Roots of Violence in Indonesia*, edited by F. Colombijn and Th. Lindblad, 33–62. Leiden: KITLV Press.

——. 2007. *Bali, an Open Fortress, 1995–2005: Regional Autonomy, Electoral Democracy and Entrenched Identities*. Singapore: NUS Press.

——. 2009. "Onafhankelijkheid of Moderniteit? Een Geïllustreerde Hypothese." In *Het Koloniale Beschavingsoffensief: Wegen naar het Nieuwe Indië*, edited by Marieke Bloembergen and Remco Raben, 105–120. Leiden: KITLV Press.

——. 2012. "Decentralization and Democracy in Indonesia: Strengthening Citizenship or Regional Elites?." In *Routledge Handbook of Southeast Asian Politics*, edited by R. Robison, 229–241. London/New York: Routledge.

Schulte Nordholt, Henk and Gerry van Klinken. eds. 2007. *Renegotiating Boundaries: Local Politics in Post-Suharto Indonesia*. Leiden: KITLV Press.

Schulte Nordholt, Henk and Margreet van Till. 1999. "Colonial Criminals in Java, 1870–1910." In *Figures of Criminality in Indonesia, the Philippines, and Colonial Vietnam*, edited by Vicente L. Rafael, 47–69. Ithaca: Southeast Asia Program.

Schulte Nordholt, Nico. 1982. "De Positie van de Lurah: Een Kritische Beschouwing naar Aanleiding van enkele Lurah-verkiezingen in Midden-Java." *Bijdragen tot de Taal-, Land- en Volkenkunde* 138: 98–128.

——. 1987. "From LSD to LKMD: Participation at the Village Level." In *Local Leadership and Programme Implementation in Indonesia*, edited by Ph. Quarles van Ufford, 47–63. Amsterdam: Free University Press.

Scott, James. 1972. "Patron-Client Politics and Political Change in Southeast Asia." *The American Political Science Review* 66: 91–113.

Scott, William Henry. 1974. *The Discovery of the Igorots: Spanish Contacts with the Pagans of Northern Luzon*. Quezon City: New Day Publishers.

——. 1981. *Boat Building and Seamanship in Classic Philippine Society*. Manila: National Museum.

——. 1983. "Oripun and Alipin in the Sixteenth Century Philippines." In *Slavery, Bondage and Dependency in Southeast Asia*, edited by Anthony Reid, 138–155. New York: St. Martin's Press.

——. 1994. *Barangay: Sixteenth-Century Philippine Culture and Society*. Quezon City: Ateneo de Manila Press.

Scott, J.G. and J.P. Hardiman. 1900–1901. *Gazetteer of Upper Burma and the Shan States*. Rangoon: Superintent of Government Printing.

Scott, Rosemary and John Guy. eds. 1995. *South East Asia & China: Art, Interaction & Commerce*. Colloquies on Art and Archaeology in Asia 17. London: Percival David Foundation of Chinese Art.

Screech, Timon. 2002. *The Lens within the Heart: The Western Scientific Gaze and Popular Imagery in Later Edo Japan*. Honolulu: University of Hawai'i Press.

Severo, Tiburcio, Florencio Asiddao and Martin Reyes. 1962. "Forest Resources Inventory in the Philippines." *Philippine Journal of Forestry* 18 (1–4): 1–19.

Sibinga Mulder. 1927. "De Economische Beteekenis van het Vee in Nederlandsch Oost-Indië en de Regeeringszorg ervoor." *De Indische Gids* 19 (1): 308–327.

Sidel, John. 1998. "Macet Total: Logistics of Circulation and Accumulation in the Demise of Indonesia's New Order." *Indonesia* 66: 159–194.

Sieh, Kerry. 2012. "The Sunda Megathrust: Past, Present and Future." In *From the Ground Up: Perspectives on Post-Tsunami and Post-conflict Aceh*, edited by Patrick Daly, R. Michael Feener and Anthony Reid, 1–22. Singapore: ISEAS.

Sieh, Kerry Danny Natawidjaja et al. 2008. "Earthquake Supercycles inferred from the Sea-level Changes Recorded in the Corals of West Sumatra," *Science* 322: 1674–1677.

Skinner, G.W. 1957. *Chinese Society in Thailand*. Ithaca, NY: Cornell University Press.

Sklair, Leslie. 2001. *The Transnational Capitalist Class*. Blackwell: Oxford.

Slater, Dan. 2010. *Ordering Power: Contentious Politics and Authoritarian Leviathans in Southeast Asia*. Cambridge: Cambridge University Press.

Smail, John R.W. 1961. "On the Possibility of an Autonomous History of Modern Southeast Asia." *Journal of Southeast Asian History* 2 (1): 72–102.

Smith, Pamela H. 1996. *Possessing Nature: Museums, Collecting and Scientific Culture in Early Modern Italy*. Berkeley: University of California Press.

——. 2010. "Vermilion, Mercury, Blood, and Lizards: Matter and Meaning in Metalworking." In *Materials and Expertise in Early Modern Europe*, edited by Ursula Klein and E.C. Spary, 29–49. Chicago: University of Chicago Press.

Smith, Paul J. 1991. *Taxing Heaven's Storehouse: Horses, Bureaucrats and the Destruction of the Sichuan Tea Industry, 1074–1224*. Cambridge, MA: Harvard University Press.

Smith, Peter C. 1978. "Crisis Mortality in the Nineteenth Century Philippines: Data from Parish Records." *Journal of Asian Studies* 38 (1): 51–76.

Smith, Pamela H. and Paula Findlen. eds. 2002. *Merchants and Marvels: Commerce, Science, and Art in Early Modern Europe*. New York: Routledge.

Stavorinus, J.S. 1797. *Reize van Zeeland over de Kaap de Goede Hoop en Batavia, naar Samarang, Macasser, Amboina, Suratte, enz.* Vol. 1. Leyden: A. en J. Honkoop.

Stevenson, Henry N.S. 1970. *The Economics of the Central Chin Tribes*. Farnborough: Gregg International. Reprint of 1943 edition.

Strassberg, Richard. ed. and trans. 2002. *A Chinese Bestiary: Strange Creatures from the Guideways through Mountains and Seas*. Berkeley: University of California Press.

Stutterheim, W.F. 1925. "Een Oorkonde op Koper uit het Singasarische." *Tijdschrift voor Indische Taal-, Land- en Volkenkunde* 65: 208–281.

Subrahmanyam, Sanjay. 1997. "Connected Histories: Notes towards a Reconfiguration of Early Modern Eurasia." *Modern Asian Studies* 31 (3): 735–762.

Suhadi, Machi and M.M. Soekarto. eds. 1986. *Laporan Penelitian Epigrafi Jawa Tengah.* Berita Penelitian Arkeologi 37. Jakarta: Pusat Penelitian Arkeologi Nasional.

Suhadi, Machi and K. Richadiana. 1997. *Laporan Penelitian Epigrafi di Wilayah Provinsi Jawa Timur 47*. Jakarta: Pusat Penelitian Arkeologi Nasional.

Sulit, Carlos. 1947. "Forestry in the Philippines during the Japanese Occupation." *Philippine Journal of Forestry* 5 (1): 22–47.

Suryadinata, Leo. ed. 2005. *Admiral Zheng He and Southeast Asia.* Singapore: Institute of Southeast Asian Studies.

Sutherland, Heather. 1983. "Power and Politics in South Sulawesi: 1860–1880." *Review of Indonesian and Malaysian Affairs* 17 (Winter/Summer): 161–207.

——. 1995. "Believing Is Seeing: Perspectives on Political Power and Economic Activities in the Malay World 1700–1940." *Journal of Southeast Asian Studies* 26 (1): 133–146.

——. 2000. "Trepang and Wangkang: The China Trade of Eighteenth-Century Makassar, c. 1720s–1840s." In *Authority and Enterprise among the Peoples of South Sulawesi*, edited by R. Tol et al, 73–94. Leiden: KITLV Press.

——. 2003. "Southeast Asian History and the Mediterranean Analogy." *Journal of Southeast Asian Studies* 34–1: 1–20.

——. 2004. "Trade, Court and Company: Makassar in the Later Seventeenth and Early Eighteenth Centuries." In *Hof en Handel: Aziatische Vorsten en de Voc 1620–1720*, edited by Elsbeth Locher-Scholten and Peter Rietbergen, 85–112. Leiden: KITLV Press.

——. 2005. "Contingent Devices." In *Locating Southeast Asia: Geographies of Knowledge and Politics of Space*, edited by Paul H. Kratoska, Remco Raben and Henk Schulte Nordholt, 20–59. Singapore: NUS Press, Leiden: KITLV Press.

——. 2011. "A Sino-Indonesian Commodity Chain: The Trade in Tortoiseshell in the Late Seventeenth and Eighteenth Centuries." In *Chinese Circulations: Capital, Commodities, and Networks in Southeast Asia*, edited by Eric Tagliacozzo and Wen-Chin Chang, 172–199. Durham, NC: Duke University Press.

——. 2013. "By the Numbers: Makassar's Trade, Centralised Statistics and Local Realities." *Masyarakat Indonesia* 39–2.

Swan, Claudia. 1995. "Ad Vivum, naer het Leven, from the Life: Defining a Mode of Representation." *Word & Image* 2: 353–372.

Tamesis, Florencio. 1948. "Philippine Forests and Forestry." *Unasylva* 2: 316–325.

Tarling, Nicholas. 2001. *Imperialism in Southeast Asia: 'A Fleeting, Passing Phase'.* London: Routledge.

Taylor, Jean G. 1983. *The Social World of Batavia: European and Eurasian in Dutch Asia.* Madison: University of Wisconsin Press.

——. 2007. "Painted Ladies of the VOC." *South African Historical Journal* 57: 47–78.

——. 2009. *The Social World of Batavia: European and Eurasian in Colonial Indonesia* Madison: University of Wisconsin Press.

Taylor, Keith W. 1983. *The Birth of Vietnam*. Berkeley: University of California Press.

Tegetmeier, W.B. and C.L. Sutherland. 1895. *Horses, Asses, Zebras, Mules, and Mule Breeding*. London: Horace Cox.

Teichman, Eric. 1922. *Travels of a Consular Officer in Eastern Tibet*. Cambridge: Cambridge University Press.

Tempany, Harold and D.H. Grist. 1958. *An Introduction to Tropical Agriculture*. London: Longmans Green.

Teston, Eugène and Maurice Percheron. 1931. *L'Indochine Moderne*. Paris: Librairie de France.

Thompson, Virginia. 1941. *Thailand, the New Siam*. New York: Institute of Pacific Relations.

Thurgood, Graham. 1999. *From Ancient Cham to Modern Dialects: Two Thousand Years of Language Contact and Change*. Honolulu: University of Hawai'i Press.

Tibbetts, G.R. 1979. *A Study of the Arabic Texts Containing Material on South-East Asia*. Leiden and London: E.J. Brill.

——. 2004. *Arab Navigation in the Indian Ocean before the Portuguese*. London and New York: Routledge.

T'ien, Ju-K'ang. 1989. "The Chinese Junk Trade: Merchants, Entrepreneurs, and Coolies 1600–1850." In *Maritime Aspects of Migration*, edited by Klaus Friedland, 381–389. Koln: Boulau.

Timberlake, Michael. ed. 1985. *Urbanization and the World-Economy*. New York: Academic Press.

Tomsa, Dirk. 2008. *Party Politics and Democratization in Indonesia: Golkar in the Post-Suharto Era*. London and New York: Routledge.

Top, Gerhard van den. 1998. *The Social Dynamics of Deforestation in the Sierra Madre, Philippines*. Leiden: Centre For Environmental Science.

Trocki, Carl. 1997. "Chinese Pioneering in Eighteenth-Century Southeast Asia." In *The Last Stand of Asian Autonomies*, edited by Anthony Reid, 83–101. London: Macmillan Press, New York: St Martin's Press.

Tucker, Richard. 1988. "The Commercial Timber Economy under Two Colonial Regimes in Asia." In *Changing Tropical Forest: Historical Perspectives on Today's Challenges in Asia, Australasia and Oceania*, edited by John Dargavel, Kay Dixon and Noel Semple, 219–229. Canberra: Centre for Resource and Environmental Studies, Australian National University.

Vel, Jacqueline. 2008. *Uma Politics: An Ethnography of Democratization in West Sumba, Indonesia, 1986–2006*. Leiden: KITLV Press.

Vera, Santiago de. 1903–1909. "Memorial to the Council by Citizens of the Filipinas Islands." In *The Philippine Islands, 1493–1803; Vol. 6*, edited by Emma Blair and Alexander Robertson, 157–233. Cleveland: A.H. Clark Co. Originally published 1586.

Vidal y Soler, Sebastian. 1874. *Memoria Sobre el Ramo de Montes en las Islas Filipinas*. Madrid: Imprenta, estereotipia y galvanoplastia de Aribau y C.

——. 1875. *Memoria de la Colección de Productos Forestales presentada por la Inspección General de Montes de Filipinas en la Exposición Universal de Filadelfia*. Manila: Imprenta de la revista mercantil de J. Loyzaga y Ca.

Viljoen, Deon. 2007. "Hendrik Swellengebel (1700–1760) and His Cape Cabinet in Holland." In *Contingent Lives: Social Identity and Material Culture in the VOC World*, edited by Nigel Worden, 450–459. Rondebosch: University of Cape Town.

Vitug, Marites Dañguilan. 1993. *The Politics of Logging: Power from the Forest*. Manila: Philippine Center For Investigative Journalism.

Vos, Reinout. 1993. *Gentle Janus, Merchant Prince: The VOC and the Tightrope of Diplomacy in the Malay World, 1740–1800*. Translated by Beverly Jackson. Leiden: KITLV Press.

Vries, Jan de. 2008. *The Industrious Revolution: Consumer Behavior and the Household Economy, 1650 to the Present*. Cambridge: Cambridge University Press.

Wade, Geoff. 1993. "On the Possible Cham Origin of the Philippine Scripts." *Journal of Southeast Asian Studies* 24 (1): 44–87.

——. 2009a. "An early Age of Commerce in Southeast Asia, 900–1300 CE." *Journal of Southeast Asian Studies* 40 (2): 221–265.

——. 2009b. "The Horse in Southeast Asia prior to 1500 CE: Some Vignettes." In *Pferde in Asien: Geschichte, Handel und Kultur*, edited by Bert Fragner, Ralph Kauz, Roderich Ptak and Angela Schottenhammer, 161–177. Vienna: Verlag der Österreichischen Akademie der Wissenschaften.

Wade, Geoff and Sun Laichen. eds. 2010. *Southeast Asia in the Fifteenth Century: The China Factor*. Singapore: NUS Press.

Wagley, Charles. 1951. "Cultural Influences on Population: A Comparison of Two Tupi Tribes." *Revista do Museu Paulista* 5: 95–104.

Wahyono, Martowikrido. 1994. "The Gold of Wonoboyo." In *Old Javanese Gold (4th–15th Century): An Archaeometrical Approach*, edited by Wilhemina H. Kal et al. Amsterdam: Royal Tropical Institute.

Walraven, Willem. 1992. *Brieven aan Familie en Vrienden 1919–1941*, edited by Robert Nieuwenhuys and Frans Schamhardt. Amsterdam: Van Oorschot.

Wang Gungwu. 1958. "The Nanhai Trade: A Study of the Early History of Chinese Trade in the South China Sea." *Journal of the Malayan Branch of the Royal Asiatic Society* 31 (June): 1–135.

——. 1991. "Merchants without Empire: the Hokkien Sojourning Communities." In *The Political Economy of Merchant Empires*, edited by James Tracy, 400–421. Cambridge: Cambridge University Press.

——. 2008. "The China Seas: Becoming an Enlarged Mediterranean." In *The East Asian Mediterranean: Maritime Crossroads of Culture, Commerce, and Human Migration*, edited by Angela Schottenhammer, 7–22. Wiesbaden: Harrassowitz Verlag.

Wang Jianping. 1996. *Concord and Conflict, the Hui Communities of Yunnan Society in a Historical Perspective*. Lund: Lund University.

Warren, James. 1985. *The Sulu Zone: The Dynamics of External Trade, Slavery, and Ethnicity in the Transformation of a Southeast Asian Maritime State*. Quezon City: New Day Publishers.

——. 2007. *The Sulu Zone 1768–1898: The Dynamics of External Trade, Slavery and Ethnicity in the Transformation of a Southeast Asian Maritime State*. Singapore: NUS Press.

Wernstedt, Frederick and J.E. Spencer. 1967. *The Philippine Island World: A Physical, Cultural and Regional Geography*. Berkeley: University of California Press.

Wheatley, Paul. 1959. "Geographical Notes on Some Commodities Involved in Sung Maritime Trade." *Journal of the Malayan Branch of the Royal Asiatic Society* 32 (2): 1–139. Singapore: Royal Asiatic Soc.

——. 1961. *The Golden Khersonese: Studies in the Historical Geography of the Malay Peninsula before A.D. 1500*. Kuala Lumpur: University of Malaya Press.

White, Benjamin. 1973. "Demand for Labour and Population Growth in Colonial Java." *Human Ecology* 1 (3): 217–236.

Whitford, Harry. 1911. *The Forests of the Philippines, Part I: Forest Types and Produce*. Manila: Bureau of Printing.

Whitney, Gordon G. 1994. *From Coastal Wilderness to Fruited Plain: A History of Environmental Change in Temperate North America from 1500 to the Present*. Cambridge: Cambridge University Press.

Wickberg, Edgar. 1965. *The Chinese in Philippine Life, 1850–1898*. New Haven/London: Yale University Press.

Wicks, R.S. 1986. "Monetary Developments in Java between the Ninth and the Sixteenth Centuries: A Numismatic Perspective." Indonesia 42: 42–77.

Wielenga, D.K. 1926. *Soemba*. 's-Gravenhage: Algemeene boekhandel voor inwendige- en uitwendige zending.

Williams, Michael. 2003. *Deforesting the Earth: From Prehistory to Global Crisis*. Chicago: University of Chicago Press.

Wilson, Chris. 2012. "Public Works and Ethnic Conflict." *Inside Indonesia*, 16 April.

Wilson, Ian. 2012. "Testing the Boundaries of the State: Gang, Militias, Vigilantes and Violent Entrepreneurs in Southeast Asia." In *Routledge Handbook of Southeast Asian Politics*, edited by R. Robison, 288–301. London/New York: Routledge.

Wilson, Lee and Eryanto Nugroho. 2012. "For the Good of the People?" *Inside Indonesia*, 19 August.

Wolf, Eric R. 1982. *Europe and the People without History*. Berkeley: University of California Press.

Wolters, O.W. 1967. *Early Indonesian Commerce: A Study of the Origins of Srivijaya*. Ithaca, NY: Cornell University Press.

——. 1982. *History, Culture and Region in Southeast Asian Perspectives*. Singapore: ISEAS.

——. 1999. *History, Culture and Region in Southeast Asian Perspectives*. Ithaca: Cornell Southeast Asian Program. Revised edition.

Worcester, Dean. 1914. *The Philippines Past and Present; Vol. 2*. London: Mills & Boon Ltd.

Worster, Donald. 1994. *Nature's Economy: A History of Ecological Ideas*. New York: Cambridge University Press.

Worsfold, W.B. 1893. *A Visit to Java, with an Account of the Founding of Singapore*. London: Bentley and Son.

Worsley, Peter. 1972. *Babad Buleleng: A Balinese Dynastic Genealogy*. Den Haag: Nijhoff.

Wrigley, E.A. 1969. *Population and History*. London: McGraw-Hill.

Xenos, Peter. 1998. "The Ilocos Coast since 1800: Population Pressure, the Ilocano Diaspora, and Multiphasic Response." In *Population and History: The Demographic Origins of the Modern Philippines*, edited by Daniel F. Doeppers and Peter Xenos, 39–70. Madison: University of Wisconsin.

Xenos, Peter and Shui-Meng Ng. 1998. "Nagcarlan, Laguna: A Nineteenth-Century Parish Demography. In *Population and History: The Demographic Origins of the Modern Philippines*, edited by Daniel F. Doeppers and Peter Xenos, 183–252. Madison: University of Wisconsin.

Yang, Bin. 2004. "Horses, Silver, and Cowries: Yunnan in Global Perspective." *Journal of World History* 15 (3): 281–322.

Yarwood, Alexander T. 1989. *Walers: Australian Horses Abroad*. Melbourne: Melbourne University Press.

Yepes, Victoria. 1996. *Una Etnografía de los Indios Bisayas del Siglo XVII*. Madrid: Consejo Superior de Investigaciones Científicas.

Younghusband, G.J. 1995. *Eighteen Hundred Miles on a Burmese tat, through Burmah, Siam and the Eastern Shan States*. New Delhi: Asian Educational Services. Reprint of 1888 edition.

Yuki Sawai et al. 2009. "A Periodic Recurrence of Geologically Recorded Tsunamis during the Past 5500 Years in Eastern Hokkaido, Japan." *Journal of Geophysical Research* 114, B01319.

Zandvliet, Kees. 2002. *The Dutch Encounter with Asia, 1600–1950*. Amsterdam: Rijksmuseum, Zwolle: Waanders.

Zeilinga de Boer, Jelle and Donald Theodore Sanders. 2002. *Volcanoes in Human History: The Far-Reaching Effects of Major Eruptions*. Princeton: Princeton University Press.

Zoetmulder, P.J. 1968. "Les grandes Religions d'Indonésie." In *Les Religions d'Indonésie*, edited by W. Stöhr and P. Zoetmulder, 257–390. Paris: Payot.

——. 1974. *Kalangwan: A Survey of Old Javanese Literature*. The Hague: Martinus Nijhoff.

——. 1982. *Old Javanese-English Dictionary*. 2 vols. The Hague: Martinus Nijhoff.

Index

www.ingramcontent.com/pod-product-compliance
Lightning Source LLC
Chambersburg PA
CBHW071227230926
51849CB00001B/45
* 9 7 8 9 0 0 4 2 8 8 0 4 1 *